绿色"一带一路"
与2030年可持续发展议程
——对接2030年可持续发展目标 促进生物多样性保护

周国梅　史育龙　[美]凯文·加拉格尔（Kevin P.Gallagher）　蓝　艳　等/著

中国环境出版集团·北京

图书在版编目（CIP）数据

绿色"一带一路"与2030年可持续发展议程：对接2030年可持续发展目标 促进生物多样性保护：汉文、英文/周国梅等著. —北京：中国环境出版集团，2021.8
中国环境与发展国际合作委员会专题政策研究报告
ISBN 978-7-5111-4729-5

Ⅰ.①绿… Ⅱ.①周… Ⅲ.①环境保护政策—研究—中国—汉、英②生物多样性—生物资源保护—研究—中国—汉、英 Ⅳ.①X012②X176

中国版本图书馆 CIP 数据核字（2021）第 095656 号

出 版 人	武德凯	
责任编辑	黄 颖	
责任校对	任 丽	
封面设计	岳 帅	

出版发行 中国环境出版集团
（100062 北京市东城区广渠门内大街 16 号）
网　　址 http://www.cesp.com.cn
电子邮箱 bjgl@cesp.com.cn
联系电话 010-67112765（编辑管理部）
　　　　　010-67162011（第四分社）
发行热线 010-67125803，010-67113405（传真）

印　　刷 北京中科印刷有限公司
经　　销 各地新华书店
版　　次 2021 年 8 月第 1 版
印　　次 2021 年 8 月第 1 次印刷
开　　本 787×1092　1/16
印　　张 12.25
字　　数 185 千字
定　　价 58.00 元

【版权所有。未经许可，请勿翻印、转载，违者必究。】
如有缺页、破损、倒装等印装质量问题，请寄回本集团更换

中国环境出版集团郑重承诺：
中国环境出版集团合作的印刷单位、材料单位均具有中国环境标志产品认证；
中国环境出版集团所有图书"禁塑"。

专题政策研究项目组成员

中国环境与发展国际合作委员会

中外组长

周国梅	生态环境部对外合作与交流中心主任
史育龙	中国城市和小城镇改革发展中心主任
Kevin P. Gallagher	波士顿大学教授，波士顿大学全球发展政策中心主任

顾　问

郭　敬	生态环境部国际合作司司长
潘家华	中国社会科学院城市发展与环境研究所所长
高吉喜	生态环境部卫星环境应用中心主任
叶燕斐	中国银保监会政策研究室主任
马克平	中国科学院生物多样性委员会副主任， 中国科学院植物研究所研究员
魏仲加	国合会外方首席顾问，国际可持续发展研究院原院长
Aban Marker Kabraji	世界自然保护联盟亚洲区办公室主任（巴基斯坦）
Guido Schmidt-Traub	联合国可持续发展解决方案网络执行主任
David Wilcove	普林斯顿大学生态学、进化生物学与公共事务教授
Thomas Lovejoy	乔治梅森大学环境工程学教授

Andrew Deutz	大自然保护协会全球政策、机构和保护金融总监
Karen Kemper	世界银行全球环境、自然资源及蓝色经济实践部主任
Caroline Kaiser	大自然保护协会自然投资机构总经理
Rose Niu	保尔森基金会自然和环境保护项目总监
Nathalie Bernasconi	国际可持续发展研究院高级总监
Jesús Ramos-Martín	厄瓜多尔伊基央亚马逊地区大学生物多样性学院院长，生态经济学家

项目组成员

葛察忠	生态环境部环境规划院环境政策部主任
蓝 艳	生态环境部对外合作与交流中心副处长、高级工程师（协调员）
Rebecca Ray	波士顿大学全球发展政策中心研究员
董 亮	外交学院亚洲研究所助理研究员
王丽霞	生态环境部卫星环境应用中心高级工程师
倪碧野	中国城市和小城镇改革发展中心高级城市规划师
彭 宁	生态环境部对外合作与交流中心副室主任
李盼文	生态环境部对外合作与交流中心工程师
张 敏	生态环境部对外合作与交流中心副室主任

本专题政策研究项目组中外组长、成员、顾问以其个人身份参加研究工作，不代表其所在单位及中国环境与发展国际合作委员会（国合会）观点。

SPECIAL POLICY STUDY TEAM MEMBERS

China Council for International Cooperation on Environment and Development

Co-Leaders:

Zhou Guomei	Chinese Co-Leader, Director General of Foreign Environmental Cooperation Center, Ministry of Ecology and Environment (MEE)
Shi Yulong	Chinese Co-Leader, Director General, China Center for Urban Development, National Development and Reform Commission (NDRC)
Kevin P. Gallagher	International Co-Leader, Professor of Global Development Policy, Director of Global Development Policy Center, Frederick S. Pardee School of Global Studies, Boston University

Senior Advisors:

Guo Jing	Director General, Department of International Cooperation, MEE
Pan Jiahua	Director General, Institute for Urban and Environmental Studies, Chinese Academy of Social Sciences
Gao Jixi	Director General, Center for Satellite Application on Ecology and Environment, MEE
Ye Yanfei	Director General, Policy Research Bureau, China Banking and Insurance Regulatory Commission (CBIRC)
Ma Keping	Secretary General of the Biodiversity Committee of the Chinese Academy of Sciences, Researcher of the Institute of Botany, Chinese Academy of Science
Scott Vaughan	CCICED International Chief Advisor, Former President and CEO of International Institute for Sustainable Development
Aban Marker Kabraji	Regional Director, Asia Regional Office of IUCN (Pakistan)
Guido Schmidt-Traub	Executive Director, Sustainable Development Solutions Network (SDSN)

David Wilcove	Professor of Ecology and Evolutionary Biology and Public Affairs, Princeton University
Thomas Lovejoy	Professor of Environmental Science, George Mason University
Andrew Deutz	Director, Global Policy, Institutions & Conservation Finance, the Nature Conservancy (TNC)
Karen Kemper	Global Director for the Environment, Natural Resources and Blue Economy Global Practice at the World Bank
Caroline Kaiser	Managing Director, Nature Vest, TNC
Rose Niu	Chief Conservation Officer, Paulson Institute
Nathalie Bernasconi	Senior Director, International Institute for Sustainable Development (IISD)
Jesús Ramos-Martín	Ecological Economist/Rector, Universidad Regional Amazónica IKIAM

Expert Team:

Ge Chazhong	Director, Environmental Strategy Institute, Chinese Academy for Environmental Planning, MEE
Lan Yan	Senior Engineer, Deputy Director, Foreign Environmental Cooperation Center, MEE (Coordinator)
Rebecca Ray	Post-Doctoral Research Fellow, Global Development Policy Center, Boston University
Dong Liang	Assistant Research Fellow, Institute of Asian Studies, China Foreign Affairs University
Wang Lixia	Senior Engineer, Center for Satellite Application on Ecology and Environment, MEE
Ni Biye	Senior Urban Planner, China Center for Urban Development, NDRC
Peng Ning	Deputy Section Chief, Foreign Environmental Cooperation Center, MEE
Li Panwen	Engineer, Foreign Environmental Cooperation Center, MEE
Zhang Min	Deputy Section Chief, Foreign Environmental Cooperation Center, MEE

The co-leaders and members of this SPS serve in their personal capacities. The views and opinions expressed in this SPS report are those of the individual experts participating in the SPS Team and do not represent those of their organizations and CCICED.

前 言

在全球抗击新冠肺炎疫情的行动中,国际社会的携手努力更加凸显了"一带一路"国际合作的重要性和必要性,共建"一带一路"合作空间进一步拓展。后疫情时代,"一带一路"倡议为世界经济增长和复苏注入强劲动力,绿色复苏亦成为疫后经济发展的重要推动力,绿色"一带一路"建设将为各国共享有韧性的、包容的、可持续的发展机制,落实联合国2030年可持续发展议程做出更加积极的贡献。

当前,生物多样性丧失和生态系统退化对人类生存和发展构成重大风险。生物多样性是人类赖以生存和发展的重要基础。2020年9月,中国国家主席习近平在联合国生物多样性峰会强调要"要同心协力,抓紧行动,在发展中保护,在保护中发展,共建万物和谐的美丽家园"。

生物多样性保护等生态环保合作一直是"一带一路"高质量发展的重要内容,联合国2030年可持续发展议程也设立了目标15(SDG15)以保护陆地生物多样性。2021年,《生物多样性公约》第十五次缔约方大会(COP15)在中国昆明召开,主题为"生态文明:共建地球生命共同体"。COP15审议《2020年后全球生物多样性框架》,确定2030年全球生物多样性保护目标,并制定2021—2030年新的十年全球生物多样性保护战略,开启2020年后全球生物多样性保护的治理进程。推动绿色"一带一路"建设与落实可持续发展目标15的协同,为共同应对全球生物多样性面临的严峻挑战、共谋全球生态文明之路、共建地球生命共同体贡献思路和解决方案。

本书以中国环境与发展国际合作委员会（以下简称国合会）绿色"一带一路"与2030年可持续发展议程专题政策研究第二期项目（2020年）研究成果为基础编辑而成，"一带一路"绿色发展国际联盟及其合作伙伴参与了该项研究。第一期项目（2018—2019年）研究提出了"一带一路"绿色发展的总体原则、目标和实施路径。在此基础上，以2021年中国举办联合国《生物多样性公约》第十五次缔约方大会为契机，第二期项目以协同推进2030年可持续发展目标为切入点，聚焦生物多样性保护，对中国及国际社会落实SDG15相关的政策标准、投资工具、治理结构以及相关实践进行了梳理和对比分析，研究提出了绿色"一带一路"建设的路线图及对接2030年可持续发展议程的政策建议。

在此，特别感谢参与研究工作的中外专家，他们是：

执行摘要：周国梅、史育龙、Kevin P. Gallagher；

第1章：周国梅、Kevin P. Gallagher、李盼文、葛察忠、Rebecca Ray；

第2章：蓝艳、李盼文、Rebecca Ray、王丽霞；

第3章：Kevin P. Gallagher、葛察忠；

第4章：彭宁、蓝艳、Rebecca Ray、董亮；

第5章：史育龙、周国梅、倪碧野、Kevin P. Gallagher。

感谢国合会首席顾问刘世锦、魏仲加对专题政策研究项目的指导，以及中外顾问对项目提出的宝贵建议。感谢国合会助理秘书长李永红，秘书处张慧勇、刘侃等对该项目的大力支持。

Forward

The global fight against COVID-19 pandemic highlights the importance and necessity of international cooperation under the framework of Belt and Road Initiative (BRI). The cooperation area of jointly building the Belt and Road is further expanded. In the post-pandemic era, BRI has injected strong momentum for global economic growth and recovery with green recovery becoming a key driver for an economic rebound. A green BRI will further contribute to a resilient, inclusive and sustainable development mechanism that could be shared by all BRI participating countries, and to the implementation of the UN 2030 Agenda for Sustainable Development.

At present, the loss of biodiversity and the degradation of the ecosystem pose a major risk to human survival and development. Biodiversity provides the very basis for the human race to survive and thrive. At the United Nations Summit on Biodiversity held in September 2020, Chinese President Xi Jinping stated that "it falls to all of us to act together and urgently to advance protection and development in parallel, so that we can turn Earth into a beautiful homeland for all creatures to live in harmony".

Ecological and environmental cooperation including biodiversity conservation is one of the most important content of high-quality development of BRI cooperation. The UN 2030 Agenda for Sustainable Development set the Sustainable Development Goal (SDG) 15 to protect the terrestrial biodiversity. The 15th meeting of the Conference of the Parties (COP 15) to the *Convention on Biological Diversity* (CBD) has been held in Kunming, China in 2021, with the theme of "Ecological Civilization: Building a Shared Future for All Life on Earth." COP 15 review *the Post-2020 Global Biodiversity Framework*, set up

2030 objectives and targets for the conservation of global biodiversity, formulate the strategy for the conservation of global biodiversity in a new decade (2021-2030), and launch the new course of post-2020 global biodiversity conservation. To align green BRI with SDG 15 provide solutions to addressing the severe challenges of global biodiversity, exploring ecological civilization and to building a shared future for all life on Earth.

This book is based on the outcomes of the second phase of the Special Policy Study (2020 SPS) on Green BRI and 2030 Agenda for Sustainable Development tasked by the China Council for International Cooperation on Environment and Development (CCICED). BRI International Green Development Coalition (BRIGC) and its partners participated in this research. The first phase was implemented from 2018 to 2019 and provided general principles, objectives and approaches towards the building of green BRI. Based on what has been achieved in phase one, the second phase of the SPS takes the opportunity of the COP 15, in 2030 aims to push forward UN SDGs in an coordinated way, with a special focus on *biodiversity conservation* (SDG 15). The research reviewed and analyzed the polices, standards, investment tools, governance structure and relevant practices in implementing SDG 15 between China and the world, proposed a roadmap for the building of green BRI and policy recommendations for aligning with 2030 Agenda for Sustainable Development.

Special acknowledgement should be given to the following experts:

Executive Summary: Zhou Guomei, Shi Yulong and Kevin P. Gallagher

Chapter 1: Zhou Guomei, Kevin P. Gallagher, Li Panwen, Ge Chazhong and Rebecca Ray

Chapter 2: Lan Yan, Li Panwen, Rebecca Ray and Wang Lixia

Chapter 3: Kevin P. Gallagher and Ge Chazhong

Chapter 4: Peng Ning, Lan Yan, Rebecca Ray and Dong Liang

Chapter 5: Shi Yulong, Zhou Guomei, Ni Biye and Kevin P. Gallagher.

We'd like to express our sincere gratitude to CCICED Chief Advisors Mr. Liu Shijin and Mr. Scott Vaughan for their guidance and to senior advisors for their valuable inputs to this SPS. Our gratitude also goes to Mr. Li Yonghong, CCICED Assistant Secretary-General, as well as Mr. Zhang Huiyong and Ms. Liu Kan and other colleagues for their great support.

目录 Contents

执行摘要 /3

1 绿色"一带一路"与2030年可持续发展议程的联系 /6

1.1 绿色"一带一路"建设进展 /6
1.2 关注 SDG15 的理由 /9
1.3 "一带一路"共建国家落实 SDG15 的进展 /10
1.4 "一带一路"倡议的惠益及生物多样性相关风险 /11
1.5 "一带一路"共建国家需要生物多样性政策 /15

2 SDG15 相关政策标准的分析 /16

2.1 中国经验的调查与评估 /16
2.2 与 SDG15 相关的国际标准 /20
2.3 中国与国际社会的异同 /29

3 SDG15 相关投资工具的分析 /31

3.1 中国经验的调查与评估 /31
3.2 国际经验的汇总与评估 /36
3.3 在国际层面利用中国实践经验的适用性 /41

4 SDG15 相关治理结构的分析 /44

4.1 中国经验的调查与评估 /44
4.2 关于生物多样性保护治理的国际经验汇总与评估 /47
4.3 中国与国际的异同 /50

5 政策建议：绿色"一带一路"建设路线图 /52

5.1 绿色"一带一路"建设路线图 /52
5.2 对接"一带一路"倡议与 SDG15 的政策建议 /54

附 录 /57

附录1：第1章的支持性证据 /57
附录2：第2章的支持性证据 /61
附录3：第3章的支持性证据 /65
附录4：第4章的支持性证据 /67

Executive Summary /75

1 LINKAGES BETWEEN THE GREEN BELT AND ROAD AND THE 2030 AGENDA FOR SUSTAINABLE DEVELOPMENT /79

1.1 Background and Progress of Building the Green Belt and Road Initiative /79
1.2 The Reason of Focus on SDG 15 /83
1.3 Progress of Countries along the Belt and Road in Implementing SDG 15 /85
1.4 Benefits and Biodiversity-Related Risks of BRI /87
1.5 The Need for Biodiversity Policy in the BRI /92

2 AN ANALYSIS OF RELEVANT POLICIES AND STANDARDS ON SDG 15 /94

2.1 Research and Evaluation of China's Experience /94
2.2 International Standards Related to SDG 15 /100
2.3 Areas of Convergence and Divergence between China and International Peers /113

3 ANALYSIS OF SDG 15 RELATED INVESTMENTS POSSIBILITIES /116

3.1 Survey and Assessment of the Chinese Experience /116

3.2 Survey and Assessment of International Practices /124
3.3 The Applicability of Chinese Domestic Experience for Biodiversity Finance with International Peers /131

4 ANALYSIS OF SDG15-RELATED GOVERNANCE STRUCTURE /134

4.1 Survey and Assessment of Practices in China /134
4.2 Survey and Assessment of International Practices /140
4.3 Areas of Convergence and Divergence between China and International Peers /144

5 POLICY RECOMMENDATION: CONSTRUCTION ROADMAP OF GREEN BELT AND ROAD /147

5.1 Roadmap for Building a Green BRI /147
5.2 Policy Instruments for Aligning the BRI with SDG 15 /152

ANNEXS /156

Annex 1: Supporting Evidence for Chapter 1 /156
Annex 2: Supporting Evidence for Chapter 2 /160
Annex 3: Evidence from Chapter 3 /166
Annex 4: Evidence from Chapter 4 /169

参考文献 /176

Part 1
中文部分

绿色"一带一路"与 2030 年可持续发展议程
——对接 2030 年可持续发展目标　促进生物多样性保护

Green BRI and 2030 Agenda for Sustainable Development
——Aligning with Sustainable Development Goal 15 to Promote Global Biodiversity Conservation

执行摘要

"一带一路"倡议秉承共商、共建、共享的原则,通过政策沟通、设施联通、贸易畅通、资金融通、民心相通,为各国共同发展和共享繁荣创造新机遇。在新冠肺炎疫情席卷全球之际,"一带一路"倡议也有了全新的、更加深远的意义。新冠肺炎疫情的应对显示出国际社会是一个不可分割的整体,需要通过共享有韧性的、包容的、可持续的发展机制和经济增长途径来进一步深化国际协同合作。"一带一路"倡议可以为满足这一需求贡献力量。

在提高"一带一路"沿线国家,乃至全球整体收入方面,"一带一路"倡议拥有巨大的潜能。世界银行研究表明,在"一带一路"倡议的帮助下,沿线国家的贸易和外商直接投资或将分别增长9.7%和7.6%,从而给"一带一路"沿线经济体带来高达3.4%的实际收入增长。"一带一路"沿线国家生活水平的提高对世界其他地区也大有裨益。世界银行的数据显示,"一带一路"倡议将带动全球收入增长2.9%。通过数据可以发现,"一带一路"倡议与跨太平洋伙伴关系形成鲜明对比,后者给成员国和世界其他地区带来的增长约为1.1%和0.4%[1]。

尽管大规模基础设施融资可以带来显著的效益,但大型基础设施项目通常也会带来包括生物多样性风险在内的一系列与可持续性相关的风险。"一带一路"倡议亦是如此。国内外的一些研究已经显示,基础设施投资带来的一些生物多样性风险很可能会出现在"一带一路"建设中。研究显示,"一带一路"倡议可能面临野生动植物生境丧失、入侵物种扩散、非法采伐、盗猎与山火发生频率上升等挑战,并因此阻碍野生动物的迁徙,增加野生动物的死亡率。修建道路、安装电力线路、建设电厂以及进行采矿活动则可能导致森林退化。因此,将生态环境风险管理纳入"一带一路"建设框架,对推动绿色"一带一路"与2030年可持续发展议程对接具有重要意义。

2019年年底,中华人民共和国主席习近平和法兰西共和国总统埃马纽埃尔·马克龙共同发布了《中法生物多样性保护和气候变化北京倡议》(以下简称

《倡议》)。该《倡议》充分展现了中国保护生物多样性的决心。在《倡议》中，中法共同呼吁：

> "在国家和国际层面，从所有公共和私人来源调动额外资源，用于适应和减缓气候变化，使资金流动符合实现温室气体低排放和气候韧性发展的路径，并用于生物多样性的养护和可持续利用、海洋养护、土地退化等；确保国际融资，特别是在基础设施领域的融资，与可持续发展目标和《巴黎协定》相符。"

为落实上述承诺，本书将对中国与国际社会在预防与缓解生物多样性风险方面的相关实践进行总结。将中国以生态保护红线政策为代表的多种实践模式，以及国际社会常用的各类方法加以调整后，可用作在绿色"一带一路"框架下落实生物多样性保护的重要举措。本书还提出了对接"一带一路"倡议与可持续发展目标和《巴黎协定》目标的总体战略原则，以及绿色"一带一路"建设的路线图。中国及共建"一带一路"国家都应该坚持将绿色作为"一带一路"建设的底色，发挥"一带一路"的"五通"功能，共同推动生态环境保护与应对气候变化等相关政策的落实，支持推动《生物多样性公约》《联合国气候变化框架公约》等环境国际公约履约进程。绿色"一带一路"建设路线图从战略层面对接了三个领域：绿色"一带一路"建设，联合国2030年可持续发展议程与共建"一带一路"国家发展目标。路线图主要包括四个方面的建议：一是加强政策沟通。把绿色发展作为共同理念，将绿色"一带一路"建设作为落实2030年可持续发展目标和推进全球环境治理变革的重要实践，充分发挥"一带一路"绿色发展国际联盟等平台作用。二是加强战略对接。建立绿色"一带一路"与联合国可持续发展议程的战略对接机制，积极推进环境政策、规划、标准和技术对接，依托"一带一路"生态环保大数据服务平台强化信息共享。三是加强项目管理。构建绿色"一带一路"项目管理机制，进一步强化项目的环境管理工作，防范"一带一路"建设生态环境风险。四是加强能力建设。共同实施绿色丝路使者计划等能力建设活动，推动绿色"一带一路"民心相通。

在上述绿色"一带一路"建设路线图的框架下，针对可持续发展目标15

（SDG15）及生物多样性保护，本专题政策研究提出了"一带一路"倡议与SDG15及《生物多样性公约》进行对接的政策建议。具体包括：

一是对接国际规则标准，鼓励采用较高环境标准。主动对接国际及国家承诺，将"一带一路"倡议与《生物多样性公约》《联合国气候变化框架公约》等国际公约进行对接。二是聚焦环境影响，实施"一带一路"项目分级分类管理。依托"一带一路"绿色发展国际联盟正在开展的《"一带一路"项目绿色发展指南》研究，推动制定"一带一路"项目分级分类指南，为共建"一带一路"国家及项目提供绿色解决方案。三是完善政策工具，防范"一带一路"建设生态环境风险。建议对"一带一路"重点行业、重点项目进行环境风险评估，建立常态化的环境风险监管机制，将环境污染、生物多样性保护、气候变化等环境因素作为评估的重要部分，充分运用绿色金融工具和环境风险分析方法，将生态保护红线作为对接"一带一路"倡议与SDG15的关键性工具。四是加强协同机制，以基于自然的解决方案促进可持续发展目标的有效衔接，发挥与SDG13气候行动等可持续发展目标的协同作用。

1 绿色"一带一路"与 2030 年可持续发展议程的联系

1.1 绿色"一带一路"建设进展

1.1.1 "一带一路"倡议背景、目的和成绩

2008 年国际金融危机以来，国际经济合作一直在聚焦发掘新增长点，探索新的经济发展模式。在这一背景下，中国提出了"一带一路"倡议，为全面解决可持续发展问题提出了中国方案。"一带一路"倡议秉持共商、共建、共享的原则，通过政策沟通、设施联通、贸易畅通、资金融通、民心相通，为各国共同发展和共享繁荣创造新机遇。随着当前全球性新冠肺炎疫情的肆虐，人们已经清楚地意识到，以"一带一路"倡议为代表的各类重大国际合作项目有助于加强全球合作，共同抗击新冠肺炎疫情，解决包括金融危机、气候变化、生物多样性丧失在内的各类全球挑战。

迄今为止，"一带一路"倡议已取得令人瞩目的成绩。2013—2019 年，中国与沿线国家货物贸易累计总额超过了 7.8 万亿美元，对沿线国家直接投资超过 1 100 亿美元，新承包工程合同额接近 8 000 亿美元[2]。世界银行研究[3]显示，"一带一路"倡议的实施，使得沿线经济体之间的贸易成本下降了 3.5%；同时，由于基础设施的外溢效应，这些沿线经济体与世界其他地区的贸易成本也下降了 2.8%。截至 2019 年 11 月，中国企业在"一带一路"沿线国家建设的境外经贸合作区，已累计投资 340 亿美元，上缴东道国税费超过 30 亿美元，为当地创造就业岗位 32 万个[4]。世界银行研究指出，"一带一路"倡议的实施可使沿线国家的收入提高 3.4%，全球收入增加达 2.9%。"一带一路"倡议已经被联合国认可为推动落实可持续发展议程的解决方案之一。

然而，"一带一路"倡议还有更大的潜力，尤其是在通过高质量基础设施投资和全球合作来支持生物多样性保护方面。2019 年 4 月，第二届"一带一路"

国际合作高峰论坛咨询委员会研究成果和建议报告（2019）中指出，"一带一路"倡议与联合国 2030 年可持续发展议程在促进合作、执行手段、举措等方面有很多共同之处，有望形成合力。

1.1.2 绿色"一带一路"建设进展

将"一带一路"打造成绿色发展之路一直是中国政府的初心和愿望，也是所有共建国家的共同需求和目标。近年来，中国以前所未有的力度推进生态文明建设。"生态优先、绿色发展"理念在全社会形成了广泛共识，经济发展正在从"先污染，后治理"的传统模式向以生态文明为导向的高质量发展转型。共建绿色"一带一路"，为中国和有关国家交流互鉴促进绿色转型、实现可持续发展搭建了平台。在六年的"一带一路"建设实践中，中国与共建"一带一路"国家在生态环境治理、生物多样性保护和应对气候变化等领域积极开展双边和区域合作，不断推动绿色"一带一路"走实走深，共同推动落实 2030 年可持续发展议程，取得了积极成效。

一是加强顶层设计，合作机制不断完善。 2015 年 3 月，国家发展改革委、外交部、商务部联合发布的《推进丝绸之路经济带和 21 世纪海上丝绸之路的愿景与行动》中明确提出，要在投资贸易中突出生态文明理念，加强生态环境、生物多样性和应对气候变化合作，共建绿色丝绸之路。2017 年，环境保护部发布《"一带一路"生态环境保护合作规划》，并联合外交部、国家发展改革委、商务部共同发布《关于推进绿色"一带一路"建设的指导意见》，明确了绿色"一带一路"建设的路线图和施工图。

随着"一带一路"建设的逐步推进，绿色"一带一路"已经得到越来越多国际合作伙伴的响应，目前，生态环境部已与共建"一带一路"国家和国际组织签署近 50 份双边和多边生态环境合作文件，并与中外合作伙伴共同成立了"一带一路"绿色发展国际联盟（以下简称联盟）。联盟由中外合作伙伴于第二届高峰论坛绿色之路分论坛上共同启动，并列为第二届高峰论坛圆桌峰会联合公报中专业领域多边合作倡议平台。联盟旨在打造一个促进实现"一带一路"绿色发展国际共识、合作与行动的多边合作倡议平台。截至 2020 年 12 月，已有来自 40 多个国家的 150 余家机构成为联盟合作伙伴，其中包括共建国家的政府部门、国际组织、智库和企业等在内的70余家外方机构。联盟建设的各项工作进入全面启动阶段，政策对话、专题伙伴关系和示范项目等活动正逐步推进，并启动了《"一带一路"绿色发展报告》《"一带一路"项目绿色发展指南》和《"一带一路"绿色发展案例研究报告》等联合研究项目。

二是丰富合作平台，合作模式更加务实。 稳步推进中柬环境合作中心、中老环境合作办公室等重点平台建设，积极推动生态环保能力建设活动和示范项目等。建立"一带一路"环境技术交流与转移中心（深圳），聚焦产业发展优势资源，促进环境技术创新发展与国际转移。这些重点平台将成为区域和国家层面推动"一带一路"生态环保合作的重要依托。现已启动"一带一路"生态环保大数据服务平台，开发并发布平台 App，完善"一张图"综合数据服务系统。该平台旨在借助"互联网+"大数据等信息技术，建设一个开放、共建、共享的生态环境信息交流平台，共享生态环保理念、法律法规与标准、环境政策和治理措施等信息。

三是深化政策沟通，绿色共识持续凝聚。 充分利用现有国际和区域合作机制，积极参与联合国环境大会、中国—中东欧国家环保合作部长会等活动，分享中国生态文明和绿色发展的理念、实践和成效。主动搭建"一带一路"绿色发展政策对话和沟通平台，举办第二届高峰论坛绿色之路分论坛，在世界环境日全球主场活动、联合国气候行动峰会、中国—东盟环境合作论坛等活动下举办绿色丝绸之路主题交流活动，并在生物多样性保护、应对气候变化、生态友好城市等领域，每年举办20余次专题研讨会，共建"一带一路"国家和地区超过800人参加交流。

四是务实合作成果，共建成效日渐显现。 绿色丝路使者计划是中国政府为提升中国与共建"一带一路"国家环境管理能力而打造的重要绿色公共产品，已为共建"一带一路"国家培训环境官员、研究学者及技术人员2 000余人次，遍布120多个国家。第二届高峰论坛成果清单中提出，未来三年将继续向共建"一带一路"国家环境部门官员提供1 500个培训名额。中国政府还与有关国家共同实施"一带一路"应对气候变化南南合作计划，提高共建"一带一路"国家应对气候变化能力，促进《巴黎协定》的落实。结合共建"一带一路"国家绿色发展现状和需求，通过低碳示范区建设和能力建设活动等方式，帮助共建"一带一路"国家提升减缓和适应气候变化的水平，推动共建国家能源转型，促进中国环保技术和标准、低碳节能和环保产品国际化。

1.2 关注 SDG15 的理由

2019年5月，生物多样性和生态系统服务政府间科学政策平台（IPBES）发布了《全球生物多样性和生态系统服务评估报告》。报告评估了过去50年生物多样性和生态系统服务对人类经济、福祉、粮食安全和生活质量的影响。评估结果显示，过去50年里，全球生物多样性的丧失速度在人类历史上前所未有。土地和海洋的利用、直接开发、气候变化、污染和外来入侵物种是造成全球生物多样性丧失的主要直接驱动因素，人口和社会文化、经济与技术、机构与治理制度等为重要间接驱动因素。迄今为止，75%的陆域环境被人类行为活动"严重改变"。由此带来的压力，使得《生物多样性公约》和《联合国气候变化框架公约》中的相关目标更加难以实现，亟须采取变革性的行动。同理，按照现在的保护速度和力度，要实现2030年可持续发展议程中的相关目标，必须采取革命性的改变。

2021年是一个重要的时间节点。《生物多样性公约》第十五次缔约方大会（COP15）于2021年在中国昆明召开，主题为"生态文明：共建地球生命共同体"。COP15审议《2020年后全球生物多样性框架》，确定2030年全球生物多样性保护目标，并制定2021—2030年新的十年全球生物多样性保护战略，开启2020年后全球生物多样性保护的治理进程。

联合国2030年可持续发展议程特别强调了生物多样性的重要作用，专门设定了目标14（SDG14，保护和可持续利用海洋和海洋资源以促进可持续发展）保护海洋生物多样性，设立了目标15（SDG15，保护、恢复和促进可持续利用陆地生态系统，可持续管理森林，防治荒漠化，制止和扭转土地退化，遏制生物多样性的丧失）保护陆地生物多样性。因此，COP15也为加速实现与生物多样性相关的可持续发展目标开启了机会之窗。

在国合会"绿色'一带一路'与2030年可持续发展议程"专题政策研究第一期项目成果的基础上，本期研究项目采取"分目标、分阶段"的方式将"一带一路"倡议与生物多样性相关的可持续发展目标结合起来。鉴于目前陆地生态系统退化情况严峻，本期研究首先聚焦于SDG15的落实，以此作为切入点，探讨如何鼓励共建国家借助"一带一路"建设更好地落实可持续发展目标，并向COP15提供政策建议。未来，可以采用相似的方法与措施将"一带一路"倡议与落实 SDG14 以及其他涉及生物多样性的可持续发展目标结合起来。

1.3 "一带一路"共建国家落实 SDG15 的进展

"一带一路"共建国家在实现 SDG15 方面缺乏进展。联合国可持续发展解决方案网络（SDSN）对 193 个国家的可持续发展目标实现情况进行了评估。SDSN 和贝塔斯曼基金会发布的《2019 年可持续发展报告》显示，气候（SDG13）和生物多样性（SDG14、SDG15）的发展趋势令人担忧。

SDSN 对 139 个共建"一带一路"国家（详见附录 1 表 A1-1）进行了可持续发展目标实现情况的评估。对于 SDG15 的落实情况，选择了 5 个指标进行评估，分别是：对生物多样性重要的陆地面积得到保护的比例（%）、对生物多样性重要的淡水水域面积得到保护的比例（%）、存活物种红色名录指数（Red List Index of Species Survival）、永久毁林比例（5 年平均）、入侵物种威胁（每百万人威胁）。

《2019 年可持续发展报告》显示，与"一带一路"经济走廊关系最密切的地理区域面临的挑战尤其严峻，包括东盟、西亚和南亚国家。相关结论将在后文进行详细讨论。具体评估结果如图 1-1 所示（详见附录 1 表 A1-2）。

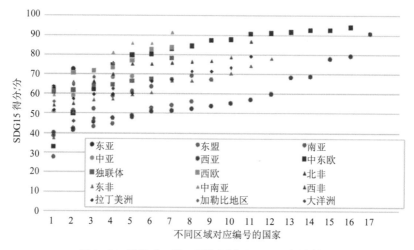

图 1-1 共建"一带一路"国家 SDG15 得分情况

注：1. 圆圈表示亚洲、方块表示欧洲、三角表示非洲、菱形表示其他区域。
2. 横坐标编号代表附录 1 的附表 A1-1 签订"一带一路"备忘录国家分布中每个区域相应编号的国家。

从 SDG15 落实状况看，在所有 139 个国家中，只有 4 个中东欧国家实现了 SDG15，分别是波兰、匈牙利、罗马尼亚和保加利亚。中东欧国家 SDG15 落实情况总体优于其他区域。其他区域国家 SDG15 的落实均存在不同程度的风险。其中，东盟成员国中，马来西亚、印度尼西亚、越南三国存在巨大风险；南亚与西亚国家中，阿富汗、伊拉克、土耳其和叙利亚四国面临重大挑战；东非国家中，吉布提、马达加斯加、塞舌尔和索马里面临重大挑战；大洋洲国家中，斐济、密克罗尼西亚、所罗门群岛和瓦努阿图面临重大挑战。

从 SDG15 落实的时序变化来看，中东欧国家同样优于其他区域。16 个中东欧国家中，有 10 个国家 SDG15 的落实进展顺利，4 个国家落实情况略有增加。东盟和南亚是 SDG15 落实情况有所下降的主要区域。其中，东盟 10 国中有 5 个国家出现下降，2 个国家工作停滞。南亚 8 个国家中有 4 个国家出现下降。中亚和独联体大部分国家工作有所停滞。其中，中亚 5 国落实情况停滞；独联体 7 个国家中有 5 个国家落实情况停滞。

从具体指标来看，对东盟和南亚国家落实 SDG15 影响最大的指标是红色名录指数，这一指数的时序变化在东盟和南亚国家都是下降的。此外，对于东盟国家，永久毁林也给 SDG15 的落实带来了巨大风险。详见附录 1。

1.4 "一带一路"倡议的惠益及生物多样性相关风险

"一带一路"倡议能够缩小基础设施发展水平的差异，加速区域一体化，促进经济发展，从而推动联合国可持续发展目标的实现。已经有证据显示，经过短短几年的发展，"一带一路"倡议已经为一些目标的落实做出了贡献。大规模开发同样也有潜在的风险，成功实施"一带一路"倡议需要实现潜在惠益最大化，并将潜在风险减至最低。其中一项风险是，如果没有充分的事前风险评估论证或在建设过程中采取风险管理措施，却在生态脆弱地区实施大型基础设施投资项目，有可能导致生物多样性减少，一旦加剧，甚至会影响到基础设施投资的经济回报。

2019 年年底，中华人民共和国主席习近平和法兰西共和国总统埃马纽埃尔·马克龙共同发布了《中法生物多样性保护和气候变化北京倡议》（以下简称《倡议》）。该《倡议》充分展现了中国保护生物多样性的决心。在《倡议》中，中法共同呼吁：

> "在国家和国际层面，从所有公共和私人来源调动额外资源，用于适应和减缓气候变化，使资金流动符合实现温室气体低排放和气候韧性发展的路径，并用于生物多样性的养护和可持续利用、海洋养护、土地退化等；确保国际融资，特别是在基础设施领域的融资，与可持续发展目标和《巴黎协定》相符。"

本报告旨在通过基于证据的研究，制定一套完整的政策框架，以帮助"一带一路"倡议对接SDG15。本部分重点阐述"一带一路"倡议的潜在和实际效益，以及在生物多样性保护方面的潜在风险。

1.4.1 "一带一路"倡议的惠益

为实现SDGs，需要投资建设必要的基础设施，到2030年，国际社会面临着相当于全球年均GDP 2.1%的融资缺口[5]。中国发起的"一带一路"倡议有可能带领全球消除融资缺口，促进实现SDGs。2019年，世界银行预计，"一带一路"倡议下的交通走廊能够提高贸易路线的速度和效率，连接交通不便的地区，并通过促进商品、服务和人员之间的流动扩大市场。基础设施项目建成后，将带来"外溢效应"，创造大量新机遇，产生新的经济活动形态。没有基础设施投资，这些效益都将无法实现[6]。

在提高"一带一路"沿线国家乃至全球整体收入方面，"一带一路"倡议有巨大潜能。世界银行研究表明，在"一带一路"倡议的帮助下，沿线国家的贸易和外商直接投资或将分别增长9.7%和7.6%，从而给"一带一路"沿线经济体带来高达3.4%的实际收入增长并带动全球收入增长2.9%。与此相比，跨太平洋伙伴关系给成员国和世界其他地区带来的增长仅为1.1%和0.4%[1]。显然，在促进参与国和全球经济繁荣方面，"一带一路"倡议拥有巨大的潜力。

"一带一路"倡议所带来的效益已经逐渐开始显现。2017年，Dreher等[7]对国家开发银行、中国进出口银行等中国金融机构在138个国家提供融资的海外项目研究发现，在项目投入使用两年后，可以带动0.7%的经济增长（平均值）。

1.4.2 生物多样性风险与"一带一路"倡议

尽管大规模基础设施融资可以带来显著的效益,但仍存在包括生物多样性风险在内的一系列与可持续性相关的风险。"一带一路"倡议亦是如此。2018年5月发表在《自然—可持续发展》(*Nature Sustainability*)上的一篇文章表示,生物多样性方面的专家指出,基础设施投资通常会带来的一些生物多样性风险很可能会出现在"一带一路"建设中,他们担心交通网络的扩张将加剧栖息地丧失、资源过度开发和周边景观的退化。[8]文章特别指出,"一带一路"倡议可能面临野生动植物生境丧失、入侵物种扩散、非法采伐、盗猎与山火发生频率上升等潜在问题,并因此阻碍野生动物迁徙,增加野生动物死亡率;修建道路、安装电力线路、建设电厂以及进行采矿活动都可能导致森林退化。作者还表示:"此类影响已经在一些地区累积到了相当高的水平。这些影响将造成生态系统服务功能退化,甚至将生态系统推至临界点,届时,任何微小的负面变化都可能导致生态系统质量和功能发生翻天覆地的变化。"[8]

一些研究已经确定了"一带一路"倡议的一些潜在生物多样性风险。2019年3月发表在《保护生物学》(*Conservation Biology*)上的一篇文章对已经提出的"一带一路"公路和铁路项目(位于"一带一路"经济走廊沿线的项目)进行了空间定位,并分析了这些项目与生物多样性关键区域(KBAs)之间的距离。

根据WWF的分析,"一带一路"经济走廊在欧亚大陆与265个受威胁物种的活动范围重合,包括39个极度濒危物种和81个濒危物种,覆盖1 739个重要鸟类保护区或生物多样性关键区域,以及46个生物多样性热点或全球200个重点生态区(Global 200 Ecoregions)。WWF发现,受影响最严重的地区是中国—中南半岛经济走廊、孟中印缅经济走廊和中蒙俄经济走廊。一份世界银行的背景研究也得出了相似的结论:中国—中南半岛经济走廊和中蒙俄经济走廊面临的森林退化导致的生物多样性丧失风险最大[10]。

为妥善应对这些风险,为"一带一路"项目提供大量必需贷款的中国金融机构制定了保障措施,与共建"一带一路"国家合作进行项目分析、评估和监管,以确保项目遵循优良实践。《自然—可持续发展》2020年的一项研究评估了与"一带一路"建设相关的金融机构的政策,包括35家中国机构和30家国际机构。研究发现，这些贷款机构中只有 17 家要求借款机构采取措施缓解对生物多样性的影响，而这其中只有一家中国机构——中国—东盟投资合作基金[11]。因此，在建立合作机制监督和缓解与"一带一路"项目相关的生物多样性风险方面，中国面临着潜在的严峻挑战。本研究将在下文详细探讨贷款机构保障措施及生物多样性风险缓解措施，以探索如何在相关领域取得进展。

生物多样性丧失也会减少经济福祉。发表在《全球环境变化》期刊上的一项研究发现，1997—2011 年，因土地覆被变化造成的生态系统服务功能损失，致使全球经济损失达每年 4 万亿~20 万亿美元[12]。世界银行 2019 年研究了肯尼亚生态保护工作的经济影响，结果表明，生物多样性管理可以影响基础设施项目周边社区使用的生态系统服务，进而对基础设施项目产生正面或负面的经济影响，造成项目结果的不同[13]。

生物多样性的风险明显对人类社区具有潜在的影响，但是这些影响对不同性别的人群可能有所不同，如果不考虑性别影响的差异，就会大幅度降低保护规划的有效性。在许多农村贫困地区，生物多样性丧失对妇女的影响远超过对男性的影响，特别是在一些地区，妇女要承担取水、收集薪柴、采摘野果的工作。这在全球发展中国家内普遍存在[14-15]。如果森林与河流生态系统遭到破坏，妇女的工作将变得更加繁重，因为她们需要在不够安全的地区内跋涉更远去完成采收工作。

除了降低妇女的生活水平之外，生物多样性丧失造成的性别差异性影响还会削弱妇女完成管理当地生物多样性这一传统任务的能力，进而对生物多样性本身造成复合型影响。在许多农村地区，妇女通过维护家庭或公共菜园来保护其社区的农业生物多样性，而男子则在农业综合企业中依靠生产单一作物获取劳动报酬[16]。在这些社区中，人们通过这两种劳动方式确保生存。在旱灾、洪灾或其他自然灾害期间，传统粮食作物品种的适应能力变得尤为重要。这些菜园还更加依赖健康的土壤和水，因为使用化学产品的成本很高。

因此，生物多样性保护支持性别平等，性别平等又会反过来支持生物多样性保护。发展项目可以支持形成良性循环，也有可能导致恶性循环，即森林与河流生物多样性丧失，同时对妇女这一农业生物多样性传统管理者产生负面影响。

相同的影响也存在于设计不当的保护项目中。在这些项目中，妇女无法再接近她们之前完成可持续采收工作而常用的森林和水道[17]。尽管项目设计者希望通过努力保护生物多样性，但是不考虑项目对不同性别群体的影响，可能会削减生物多样性的整体惠益，因为妇女必须将管理农业生物多样性的时间用于长途跋涉以便收集家用所需的基本物品。

遗憾的是，如果不是特别留意，规划者几乎很难发现这些生物多样性管理中存在的基于性别的风险。正如 Lu 等[18]在 2018 年所述，在妇女通常不参与公开讨论的环境中，发展项目对她们造成的影响将无人察觉。在这样的环境中，即使是依赖社区参与的项目也会忽视妇女的贡献，除非特别强调妇女的参与。因此，项目规划者与投资方容易面临生物多样性丧失的风险[19-21]。

如果"一带一路"倡议不能制定一系列战略性政策和标准，并形成相应机制来减少生物多样性风险，就将进一步面临金融、社会、环境和政治方面的风险，从而使其无法释放全部潜能。脆弱的生态系统可能影响基础设施项目的完整性，降低财务收益率，同时还会通过债务增加项目所在国的宏观经济压力，影响中国金融机构的资产负债表。此外，生物多样性加速退化可能导致社会冲突，造成信誉危机，进而威胁对"一带一路"倡议至关重要的地缘政治关系。出于上述原因，控制与"一带一路"倡议相关的生物多样性风险至关重要。

1.5 "一带一路"共建国家需要生物多样性政策

制定切实政策进行生态保护将是最大化激发"一带一路"潜能的关键，为最大化发挥"一带一路"倡议的积极影响，本书将对中国和世界在项目融资和生物多样性方面的最佳实践进行研究，从而为中国积累经验，在未来推动"一带一路"倡议继续发展的过程中制定一套一致的政策。本书第 2 章对中国和国际机构的生物多样性保护政策与标准进行总结和评估；第 3 章对中国和国际的生物多样性融资政策进行分析；第 4 章主要探讨可以利用哪些中国和国际治理结构将生物多样性整合进"一带一路"倡议；第 5 章将分析中国可以采取哪些政策建议对接"一带一路"倡议和 SDG15。

2 SDG15 相关政策标准的分析

2.1 中国经验的调查与评估

2.1.1 中国生物多样性保护现状

中国是世界上生物多样性最丰富的国家之一,也是生物多样性受威胁最严重的国家之一。中国加强生物多样性保护,积极开展生物多样性调查、生态系统和物种濒危等级评价、就地保护和迁地保护以及制定相应的生物多样性保护政策等工作。中国落实 SDG15 的进展评估见附录 2 表 A2-1。

在就地和迁地保护方面,中国建立了以国家公园为主体,涵盖自然保护区、风景名胜区、森林公园、地质公园、湿地公园、文化自然遗产等的自然保护地体系,并建立重点生态功能区、生物多样性保护优先区作为其重要补充。国家公园、自然保护区、森林公园、风景名胜区、地质公园、湿地公园、饮用水水源地等保护地数量达 10 000 多处,约占陆地国土面积的 18%。同时,中国也已提出全国重点生态功能区、重要生态功能区、生态敏感区和脆弱区等大尺度的生态功能区域,这些不同尺度的保护区域对中国的国土生态保护起到了积极作用。尽管进行了大规模的保护,但因保护地存在空间界限不清、交叉重叠等问题,近年来生态空间被挤占、生态系统退化严重、生物多样性加速下降的总体趋势仍在持续。划定生态保护红线,就是要明确生态空间范围内具有特殊重要生态功能、必须强制性严格保护的区域,实现一条红线管控所有重要生态空间。

2.1.2 中国生态保护红线实践

1. 生态保护红线划定和管理进程

2011 年 10 月,《国务院关于加强环境保护重点工作的意见》发布,首次提出划定生态红线,明确在重点生态功能区、陆地和海洋生态环境敏感区及脆弱区划定生态红线。2017年2月，中共中央办公厅、国务院办公厅印发《关于划定并严守生态保护红线的若干意见》，确定了生态保护红线划定与严守的指导思想、基本原则和总体目标，标志着中国生态保护红线的发展进入了全新的快速发展阶段。

2. 制定了科学的划定方法

生态保护红线划定首先需要开展科学评估，识别水源涵养、生物多样性保护、水土保持等生态功能极重要区域，以及水土流失、土地沙化、盐渍化等生态环境极敏感脆弱区域的空间分布。其次，将上述两类区域进行空间叠加，划入生态保护红线，涵盖所有国家级和省级禁止开发区域，以及有必要严格保护的其他各类保护地等。

生态保护红线的设计是为了保护几乎所有稀有的中国濒危物种及其栖息地，是结合中国环境保护管理工作的实际提出的。生态保护红线不等同于重新划定新的保护地，而是通过更加科学、全面和系统的方法，实现大尺度生态保护体系的构建与优化。生态保护红线将已有的重要保护地整合为完整且便于管理的生态保护体系，既包含已建的各类保护地，也包含现有的保护空缺区域。

3. 制定了权责分明的划定与严守体系

中国的生态保护红线划定是国家制定技术指南指导各省级政府划定，最终由省级政府自主决定具体划定范围。国家从宏观角度制定《生态保护红线管理办法》，各省根据国家制定的办法，结合本省实际，制定符合地方实际的《生态保护红线管理办法》，分别在环境准入、资源可持续利用、生态保护修复、生态保护补偿、评估考核等方面细化。生态保护红线管理和监督的责任也由各级政府来承担。

4. 形成了显著的保护成效

2018年1月，国务院批准了京津冀、长江经济带省（市）和宁夏等15个省级行政区的《生态保护红线划定方案》，且15个省级行政区均已发布实施。2018年10月，生态环境部会同自然资源部组织召开审核会议，原则通过另外16个省级行政区的生态保护红线划定方案。目前，生态保护红线具体划定范围还有待通过勘界定标等落实到具体地块，但就初步划定方案来看，全国生态保护红线划定范围约占国土面积的1/3，但红线内的林地、草地、湿地等主要生态用地面积占全国主要生态用地面积的 55%，以国家公园为主体的自然保护地体系面积占陆地国土面积的 18%以上，提前实现了 2020 年爱知生物多样性目标提出的保护地面积达到 17%的目标。大熊猫、朱鹮、藏羚羊等部分珍稀濒危物种，野外种群数量稳中有升。红线内主要生态用地涵盖了长江、黄河、珠江等Ⅲ级以上主要河流的集水区，以及所有国家级和绝大部分地方级生物多样性丰富地区，保护了绝大多数河流型和湖泊型水源地及部分地下水水源地，并保护了所有《国家重点保护野生动植物名录》中物种的分布区域以及保护动物和植物的集中分布区域。

2.1.3 中国生态保护红线政策对生物多样性保护的经验

中国生态保护红线政策对生物多样性的保护主要是将生物多样性丰富和重要地区划入生态保护红线，然后维护和修复生态保护红线内的生态环境，实现生物多样性的就地和迁地保护。

1. 生态保护红线范围科学合理

生态保护红线统筹考虑自然生态的整体性和系统性，开展科学评估，按生态功能重要性、生态环境敏感性与脆弱性划定并落实到国土空间。生态保护红线的范围涵盖所有国家级、省级禁止开发区域，以及有必要严格保护的其他各类保护地等，实现了一条红线管控重要生态空间。

2. 对生态保护红线区内人类活动进行严格管控

生态保护红线，从功能定位看，对维持生态平衡、支撑经济社会可持续发展意义重大；从用地性质看，是具有重要生态功能的生态用地，必须严格用途管制；从保护要求看，是保障和维护生态安全的临界值和最基本要求，是保护生物多样性，维持关键物种、生态系统存续的最小面积，确保功能不降低、面积不减少、性质不改变。生态保护红线原则上按禁止开发区域的要求进行管理，严禁不符合主体功能定位的各类开发活动。

（1）对生态保护红线内的国家公园、自然保护区、风景名胜区、森林公园、地质公园、世界自然遗产、湿地公园、饮用水水源保护区等各类保护地的管理，法律法规和规章另有规定的，从其规定。

（2）红线内其他区域，制定了生态保护红线内禁止开展的人类活动类型，如矿产资源开发活动；围填海、采砂等破坏海河湖岸线等活动；大规模农业开发活动，包括大面积开荒，规模化养殖、捕捞活动，纺织印染、制革、造纸印刷、石化等制造业活动，房地产开发活动，客（货）运车站、港口、机场建设活动，火力发电、核力发电活动，以及危险品仓储活动等；生产《环境保护综合名录（2017年版）》所列"高污染、高环境风险"产品的活动；《环境污染强制责任保险管理办法》所指的环境高风险生产经营活动。

3. 对生态保护红线区内实行生态修复和生态补偿

（1）开展生态修复

制定生态保护红线保护与修复方案，优先保护良好生态系统和重要物种栖息地，修复受损生态系统，构建生态廊道和重要生态节点，提高生态系统完整性和连通性。将生态保护红线保护修复作为山水林田湖草沙等各类生态保护修复工程的重要内容，统筹生态保护红线内水土保持、天然林资源保护、国土综合整治等各类生态保护修复工程资金渠道，落实保护与修复资金。按照陆海统筹、综合治理的原则，开展海洋国土空间生态保护红线的生态整治修复，重点加强生态保护红线内入海河口、滨岸带、海岛和受污染海域综合整治。

（2）开展政府投入和多元补偿的生态补偿机制

各级人民政府加大对生态保护红线所在地区财政资金投入力度，鼓励各地出台有利于生态保护红线的财政、信贷、金融、税收等政策，建立生态补偿机制。

地方各级人民政府应建立政府引导、市场运作、社会参与的多元化投融资机制，引导社会力量参与生态系统保护与修复，鼓励在生态保护红线内开展生态系统服务付费试点，探索生态产品价值实现的市场化机制。

4. 建设和完善生态保护红线综合监测网络体系

及时获取监测数据，加强监测数据集成分析和综合应用，全面掌握生态保护红线生态系统构成、分布与动态变化，实时监控人类干扰活动。提高管理决策科学化水平，及时核查和处理违法行为。

5. 创建生态保护红线严守责任体系

（1）强化执法监督

建立生态保护红线常态化执法机制，定期开展执法督查，及时发现和依法处罚破坏生态保护红线的违法行为，切实做到有案必查、违法必究。

（2）建立考核机制

开展生态保护红线保护成效考核，并将考核结果纳入生态文明建设目标评价考核体系，作为对地方政府工作成效进行评判的重要参考。

（3）强化责任追究

严格领导干部责任追究，尤其是对造成生态环境和资源严重破坏的，要实行终身追责，责任人无论是否已调离、提拔或退休，都必须严格追责。

（4）创建激励约束机制

对生态保护红线保护成效突出的单位和个人予以奖励，并提出根据需要设置生态保护红线管护岗位，提高居民参与生态保护积极性。

（5）扩大公众知情和参与

及时准确发布生态保护红线分布、调整、保护状况等信息，保障公众知情权、参与权和监督权。加大政策宣传力度，发挥媒体、公益组织和志愿者作用。

2.2　与SDG15相关的国际标准

在建设"一带一路"过程中，针对SDG15需求采取一系列协调统一的标准，不仅有助于将风险降至最低，同时能够最大限度地发挥"一带一路"倡议的积极作用，确保参与各方的行为合法合规。实现这一目标的主要方法是加强环境与社会风险管理（ESRM）。本节将回顾全球主要多边金融机构在基础设施、一体化和发展金融方面所践行的与SDG15相关的国际标准。本节由两部分内容组成，第一部分简要概述制定绿色"一带一路"标准的益处，第二部分对国际参与者采取的主要政策进行对比分析。

在过去的几十年中，国际金融和投资领域越来越广泛地采用了环境评估和监督体系。本节根据"一带一路"项目融资领域最为活跃的中国金融机构，选取了相应的国际金融机构，并对其惯常做法进行了汇总分析。"一带一路"项目主要通过丝路基金、国家开发银行、中国进出口银行等中国官方机构获得融资，但其资金渠道并不仅限于此[22]。因此，在探讨跨国基础设施建设项目中的环境治理问题时，本研究选择了经常为共建"一带一路"国家提供资金支持的传统多边金融机构（DFIs）作为研究对象。

2.2.1 制定绿色"一带一路"标准和保障措施的益处

研究制定绿色"一带一路"标准不仅能确保"一带一路"倡议与SDGs对接，还有利于保证各利益相关方获益。因此，高水平或最佳的环境标准要充分考虑参与"一带一路"建设的中国和其他利益相关方的诉求，从而保证"一带一路"倡议通过提供公共产品，造福全球经济。

表2-1 在"一带一路"倡议中实现SDGs标准化的益处

中国参与者	扩大市场 提升项目效果 防止违约风险 预防与缓解环境和社会风险 预防与缓解信誉风险
项目所在国	改善财政资源管理 加强自然资源管理 提高机构能力 预防与缓解环境和社会风险 预防与管理信誉风险
当地社区	降低出现社会冲突的可能性 促进当地社区发声，提高项目归属感 降低脆弱性 提高生活水平
全球	平等利用资源 提升全球公共产品供给 互联互通与全球增长 领导力与合法性

资料来源：在世界银行（2010）和国家开发银行—UNDP（2019）研究基础上总结。

绿色"一带一路"标准还能提升项目业绩和收益率。2018年，国际金融公司（IFC）发现，围绕上述提到的各项共同准则制定的标准与良好的金融业绩（以资产收益率和净资产收益率计算）和金融风险评级之间存在关联性。该研究涵盖656个IFC项目，总价值达到370亿美元。基于债务可持续性分析（DSA）的风险工具有助于确保中国参与者免于承担项目债务违约的风险。尽管很难以量化方式评估环境和社会风险管理（ESRM）的成本和效益，2010年，世界银行独立评估局（IEG）（一家独立监督机构）还是对 ESRM 的成本与效益进行了评估，并得出结论：环境保障的效益远远超过新增成本[23]。在选取银行项目作为样本进行风险与收益评估后，世界银行发现，大多数敏感项目"对项目所在国来说，要么是低成本低效益，要么是高成本高效益"。通过上述同一个 IEG 调研与评估研究，世界银行还发现，参与调查的工作组组长中，"超过一半表示银行的保障提高了受益方对项目的接受度，保障政策也让将近 30% 的联合投资者增加了对项目的接受度"[23]。

专栏 2-1　案例研究：在秘鲁的中国矿产企业纳入 ESRM

中国金融机构、企业和政府可以围绕共同准则，制定一套相应的标准，并从中受益。首先，这些工具可以帮助中国的银行和企业提高和保持海外市场占有率。中国在秘鲁的经验就是一个很好的例子。由于中国投资者和秘鲁政府在 ESRM 方面的工作缺失，中国在秘鲁的首个"一带一路"项目耗资巨大，成效甚微。中国企业与当地工人和社区在工人健康与安全、应急准备和生物多样性等问题上无法达成一致，虽然其中一些问题是受到项目所在国执行能力的限制，而非中国企业的过错，但从整体上看，中国企业的声誉受到了严重损害。事实上，由于当地普遍认为中国企业和金融机构缺乏有效的风险管理策略，中国企业很难在秘鲁矿产和资源开采项目中中标。此后，中国铜企在项目设计阶段就加强了环境和社会风险管理，引入了利益相关方磋商机制。这类举措使中国企业打入了秘鲁市场，并提高了中企在当地的声誉。在出现问题的时候，环境和社会风险管理计划使企业和项目所在国可以及时采取应对措施，将损失降至最低[24-25]。

标准还可以造福项目附近的当地社区。在项目建设开始前就与当地社区进行沟通，有助于找到可能出现的问题。在玻利维亚，中国锡企提前与项目拟建地社区进行沟通，了解到其对项目不完全支持。随后，玻利维亚政府找到了另一处项目建设地，当地社区更适合项目实施，准备也更充分，从而阻止了一场可能出现的社会冲突。此类冲突将影响企业的商业前景甚至损害中企的声誉[25]。

专栏 2-2　金融机构之外：联合国系统内的环境治理体系

　　通过联合国机制，各国建立了与本研究分析的金融机构治理系统相似的体系。在这方面，《生物多样性公约》长期以来一直是一个强化与协调各国家标准的全球性平台。《生物多样性公约》指南鼓励各国在信息共享和能力建设方面进行合作，以制定自己的标准和做法（《生物多样性公约》，1992，第十四条），因此与绿色"一带一路"框架高度兼容。

　　2006年，《生物多样性公约》制定了包括生物多样性在内的环境影响评估自愿准则，包括在上游阶段关注和确认需要着重注意的潜在问题领域。该准则鼓励各方在提议项目之前将重点放在上游阶段的工作上，即调查和确认生物多样性资源，例如中国近期在划定保护优先区时开展的相关工作。随后，初评和筛选各项目建议书，以确保在评估阶段能够妥善应对所有可能的风险。在开展影响评估时，应尽可能确保所有利益相关方的充分参与。在对每个独立项目进行影响评估之后，应建立问责机制用于监测和管理项目风险、监督实施所有必要的缓解措施。《生物多样性公约》还呼吁协调各生物多样性融资机制之间的标准，列出了适用于所有情况的标准，包括但不限于：重视并优先考虑生物多样性的内在价值及其在当地生计中的作用；项目利益相关方的有效参与；建立体制框架以监督保障措施的实施。

　　全球环境基金（GEF）已经成为环境标准指南的另一个重要来源。全球环境基金不单独为项目提供资金，而是通过共同筹资的方式开展工作。因此，GEF标准可以"挤进"其他贷方标准并扩大覆盖范围。全球环境基金对项目有9项最低标准，包括评估、问责机制，保护措施，对土地使用和现有社区非自愿迁居的限制等。第一项最低标准涉及环境和社会评估、管理和监测，与《生物多样性公约》的指导原则相呼应，此标准要求尽早进行项目初审与筛选，以确定哪些风险（既包括本项标准涉及的风险，也包括其余8项标准下涉及的风险）属于各项目共有的风险。第二项标准要求建立如下文所述的体制机制，以解决那些可能以可追责、透明方式出现的问题。尽管这些保障措施的范围代表了国际开发金融环境管理中的关键要素，但其规模并不大。全球环境基金目前四年工作周期内只募集到了41亿美元的认捐资金，在主要金融机构提供的发展资金中，这仅仅是一小部分。相比而言，世界银行在过去四年中批准的项目金额超过了1 200亿美元[26]。因此，本研究关于国际经验的探讨将重点放在最大的金融机构（通常是为发展中国家提供基础设施融资的机构）上，作为与"一带一路"项目进行比较的切入点。

> **专栏 2-3　金融机构之外：私营部门的环境治理体系**
>
> 除了本章介绍的多边方法外，近年来以私人投资与金融为主导的体系建设也取得了重大进展。其中最为人所熟知的当属《赤道原则》，专门用于支持私营金融机构开展项目建议书评估工作。《赤道原则》着重于项目的早期审查和分类，以确保项目层面的评估能够妥善地解决所有重大环境和社会风险，同时确保尽可能广泛的公众参与。《赤道原则》强调设计完善的机构问责机制，并与国家司法补偿措施结合使用，以确保在实践中进行适当的项目管理（《赤道原则》，2020）。作为《赤道原则》的补充，国际标准化组织建立了 ISO 14000 环境管理系列标准，作为环境管理工具。这些体系并未指定具体的保障措施，但划定了各机构建立各自标准时的范围，并承诺对员工进行培训和审核以确保合规。
>
> 虽然这些框架可以作为重要工具帮助私人贷方和投资者更好地选择和管理项目，但其与"一带一路"投资项目并不完全相同，因为"一带一路"项目涉及各国政府之间的合作。因此，本节将重点介绍经常为发展中国家提供基础设施融资的金融机构的常规做法。

2.2.2　国际金融机构生物多样性相关政策与措施的对比分析

本节针对 11 家在全球范围内开展基础设施融资服务的主要国际金融机构进行了研究，重点关注其与生物多样性相关的实践措施。分析结果表明，这些机构的目标和指导原则高度一致，几乎所有机构都寻求将生物多样性风险降至最低，在生物多样性方面实现净零损失甚至净收益整体目标。此外，大多数机构还要求将生物多样性评估与缓解措施绑定，并将利益相关方参与和磋商纳入生物多样性评估与管理。对具体操作与政策的详细分析也显示，各机构之间存在明显的共性。

大多数国际金融机构将生物多样性保护目标作为其工作的核心。亚洲基础设施投资银行（AIIB）、拉丁美洲开发银行（CAF）以及世界银行（WB）和国际金融公司（IFC）都已经认识到，有必要"通过可持续利用生物多样性和自然资源在经济、社会和文化方面的多重价值，将保护需求和开发重点进行整合"。为具体落实这些目标，上述金融机构采取生物多样性"净零损失"政策（如 AIIB）或"净零损失或净收益"政策［如欧洲投资银行（EIB）、亚洲开发银行（ADB）、德国复兴信贷银行（KFW）和拉丁美洲开发银行］。

在生物多样性保护的总体原则和政策措施方面，大多数金融机构的实践也高度相似。几乎所有的机构都要求满足以下五个条件：

- 对接国际承诺和国家法律要求；
- 根据生物多样性要求制定禁止项目类别清单；
- 要求开展生物多样性评估和环境影响评估；
- 依照缓解措施层级（mitigation hierarchy）采取措施，确保生物多样性净零损失或净收益；
- 在生物多样性评估与管理过程中确保利益相关方的参与和磋商。

这些政策详见表 2-2 和表 A2-2（见附录 2）。在表 2-2 中，表纵向列出了金融机构名称，横向列出了具体的生物多样性措施。应该注意到，尽管这些机构制定了相关政策，却并不能保证其随时得到落实，从而对项目、生物多样性和社区造成负面影响[27]。

表 2-2　金融机构在生物多样性保护方面的操作要求

	生物多样性保护的国际最佳实践				
	对接国际承诺和国家法律要求	禁止项目类别的排除清单	生物多样性影响评估	依照缓解措施层级采取措施	利益相关方参与和磋商
ADB	X	X	X		X
AFDB	X	X	X	X	X
AIIB	X	X	X	X	X
BNDES	X		X	X	
CAF	X	X	X		X
EBRD	X	X	X	X	X
EIB	X	X	X	X	X
IDB	X	X	X	X	X
IFC	X	X	X	X	X
KFW	X	X	X		X
WB	X	X	X	X	X

注：ADB：亚洲开发银行；AFDB：非洲开发银行；AIIB：亚洲基础设施投资银行；BNDES：巴西国家开发银行；CAF：拉丁美洲开发银行；EBRD：欧洲复兴开发银行；EIB：欧洲投资银行；IDB：美洲开发银行；IFC：国际金融公司；KFW：德国复兴信贷银行；WB：世界银行。
资料来源：作者对官方文件与采访的分析汇总。

（1）对接国际承诺和国家法律要求

确保机构的实践符合具体的国际或国家承诺和法律要求是所有国际机构的一个共同特征。以亚洲基础设施投资银行（以下简称亚投行）为例，"亚投行不会有意地资助下述有关项目……依据项目所在国的法律法规或国际公约和协议，被判定为非法或在国际上逐步被淘汰或禁止产品的生产和交易或相关活动"[28]。研究关注的大部分机构都有类似的表述。亚投行和这些机构提供了一份说明性清单，列出了其所遵守的所有国际和国家承诺（详见下文"排除清单"）。

（2）禁止项目类别的排除清单

亚投行和其他国际机构提供了一份说明性清单，列出了其所遵守的所有国际和国家承诺，大多都是对接性表述。以亚投行为例，清单包括如下内容[28]：

- 受《濒危野生动植物种国际贸易公约》（CITES）管制的野生动植物贸易或野生动植物产品的生产或贸易。
- 项目所在国的法律，或与保护生物多样性资源或文化资源有关的国际公约，如《波恩公约》《拉姆萨尔公约》《世界遗产公约》《生物多样性公约》等所禁止的活动。
- 商业性采伐作业，或购买用于采伐原始热带雨林或老龄林的伐木设备。
- 来源并非可持续管理林区的木材，或其他林业产品的生产或贸易。
- 会伤害大量脆弱、受保护物种且会破坏海洋生物多样性与海洋生境的各类海洋和沿海捕捞作业，如大规模的中上层流网捕鱼和细网捕鱼。

研究关注的国际机构中，大部分都将项目排除范围进行了扩展，不仅违反各项国际和国家承诺的项目不会得到支持，环境影响评估与相关评价结果认为影响严重的项目也会被拒。大多数机构都有类似的规定，例如，非洲开发银行政策规定，"如果我行发现某一投资对环境或社会造成的不利影响无法得到妥善解决，则我行可选择中止投资……如果某一项目对动植物生境/生物多样性的影响非常严重，我行可选择停止资助该项目[29]。"

（3）生物多样性影响评估

研究关注的主要国际机构中，大多数都将"生物多样性影响"评价纳入了广义的"环境影响评估"开展相关分析。根据相关政策要求，各机构须充分考虑其项目对动植物生境和生物多样性产生的直接、间接和累积影响。世界银行会检查对生物多样性的威胁因素，例如，动植物生境减少、退化和破碎化、外来物种入侵、过度开发、水文变化、养分负荷、污染及其附带的风险、预计的气候变化等。世界银行在为生物多样性和动植物生境的重要性划分等级时，主要考虑其在全球、地区或国家层面的脆弱性和不可替代性，同时，也会把受到项目影响的各方及其他利益相关方对生物多样性和动植物生境的重视程度考虑在内[30]。本节研究的大多数机构制定的政策中都有相同或类似的表述（表2-2）。

拉丁美洲开发银行的表述和范围则略有不同。其政策指出，该机构将评估"所研究区域内的相关自然、生态和社会经济条件，尤其是可能受到拟议的发展规划明显影响的环境条件，特别是人口、动植物、土壤、水、空气、气候要素，建筑和考古遗产及景观等有形资产，以及以上多者之间的关系；研究在受项目影响的范围内正在进行或拟议的开发活动，包括与项目非直接相关的活动。[31]"

美洲开发银行（IDB）支持的项目多位于生物多样性丰富地区，如横跨多个国家的亚马孙盆地。因此，美洲开发银行的政策也涉及与项目相关的跨境生物多样性问题。美洲开发银行的环境评估力求在项目初期发现可能涉及的跨境问题，评估过程主要针对可能产生重大的跨境环境和社会影响的活动，例如，项目活动会影响其他国家使用水路、集水区、海洋/沿海资源、生态廊道、区域性空气流域和蓄水层等；该评估过程解决的问题包括：向受影响的国家通报重大的跨境影响；采取适宜措施征询受影响各方的意见；采取令IDB满意的环境缓解和/或监测措施。

除了评估生物多样性影响之外，国际机构还建议对经济影响进行性别区分，以便评估项目对妇女进行农业生物多样性管理工作的间接影响。《生物多样性公约》下《2015—2020年性别平等行动计划》呼吁，在计算项目成本与收益时，应像绿色气候基金和气候投资基金一样，针对男女采取不同的计算方法，而不是进行总体估算[32-34]。

（4）依照缓解措施层级采取措施

就强制性生物多样性影响评估而言，表2-2显示，大多数主要国际金融机构（研究涵盖的11个机构中的8个）要求遵循缓解措施层级以实现生物多样性净零损失或净收益的整体目标。缓解措施层级包含以下四个支柱。

- 避免：采取措施从源头避免产生影响，如合理安置基础设施要素，以绝对避免对生物多样性的某些组成部分产生影响。

- 最小化：对于无法绝对避免的影响，在切实可行的范围内采取措施降低影响的持续时间、强度和/或范围（包括直接、间接和累积影响）。
- 恢复/修复：在项目产生无法绝对避免的影响和/或影响已无法减至最低后，采取措施恢复退化的生态系统或修复清理后的生态系统。
- 补偿：对于无法绝对避免、缓解和/或恢复或修复的明显残留的负面影响，采取抵消补偿等措施来应对。抵消补偿措施可以是采取主动的管理干预措施，如修复退化的生境、抑制退化状态或规避风险、保护生物多样性即将或可能受到威胁的地区等。

（5）利益相关方参与和磋商

在生物多样性评估和管理过程中，本研究关注的所有机构均要求利益相关方的参与和磋商。各机构都做出了一定的承诺，将征询受影响人群和社区的意见，以确保他们知情并参与到项目进程中来。

如前所述，在聚集了大量脆弱群体的地区，人们对于生物多样性问题忧心忡忡，在这些地区进行的大型基础设施项目中，CAF的参与度大概是最高的。CAF要求在环境评估过程中尽早与受项目影响的群体进行沟通，并要求将这种沟通延续到整个项目周期。在整个项目周期中，重要信息应及时向受影响的群体、民间社会组织和其他主要利益攸关者披露。CAF还要求："应将项目对森林和自然生境的潜在影响，以及其为社区福利而获取和使用资源的权利作为'环境和社会评估'的一部分来进行评价"（CAF，2015：64）。

国际金融公司要求借款人必须执行"利益相关方参与计划"。在适用的情况下，"利益相关方参与计划"包含差异化举措，以确保弱势群体或脆弱群体能够有效地参与进来。如果利益相关方的参与主要依靠社区代表，那么客户须尽一切努力核实这些人能否切实代表受影响的社区，以及他们能否把沟通的结果如实地反馈给社区成员。如果受到影响的社区将承受项目带来的风险和不利影响，则磋商过程应当确保受影响的社区有机会就项目风险、影响和缓解措施等发表意见，客户也可考虑这些意见并做出回应[35]。

金融机构非常清楚，在其"利益相关方参与计划"中保证妇女参与非常重要，尤其是在存在社区搬迁可能性的情况下。如1.4.2"生物多样性风险与'一带一路'倡议"部分所述,在世界上许多农村贫困地区,妇女通常不参加公开讨论,却首当其冲地遭受了生物多样性损失的影响,这会削弱妇女管理农业生物多样性的能力,可能进一步加大生物多样性损失,并限制了保护项目的惠益。例如,由亚投行、亚开行、非洲开发银行、欧洲复兴开发银行、欧洲投资银行、美洲开发银行、新开发银行(NOB)和世界银行的代表组成的银行间工作组最近就有效利益相关方参与发布了联合建议,鼓励项目规划者在进行具体流程设计时优先考虑妇女和其他弱势群体的参与,并在必要时按性别分别制定利益相关方参与过程[36]。

2.3 中国与国际社会的异同

中国在实现与生物多样性相关的 SDGs 方面已经取得了显著进展,包括 SDG15。《中国履行〈生物多样性公约〉第六次国家报告》显示,中国在大部分领域取得了令人满意的成绩,其中一些领域与"一带一路"倡议关系密切,包括将生物多样性融入国家和地方规划、动员和增加生物多样性融资,以及保护内陆淡水、森林和山地生态系统。在上述三个领域,中国与其他国家已经形成合力,有助于促进"一带一路"国际合作。

在取得积极进展的过程中,中国采用的一项重要机制就是利用生态保护红线对生态脆弱区和重要生态功能区进行保护。2017 年,中共中央办公厅、国务院办公厅联合确立了生态保护红线的基本框架,生态保护红线迅速在全国推广开来。2018 年,国务院批准了京津冀、长江经济带省(市)和宁夏等 15 个省级行政区的生态保护红线划定方案。到 2020 年年底,其他省级行政区将全部完成生态保护红线划定工作,按计划,全国生态保护红线约占国土面积的 1/3。

国际上,生物多样性保护主要围绕以下方面展开:建立和统一全球标准,在划定保护区的同时,制定可操作的风险管理策略保护当地生态系统、维护机构声誉和参与项目的所有伙伴之间的合作关系。包括多边和国家开发银行在内的主要国际金融机构使用五大方法实施国际标准:①在实践过程中与国际或国家承诺对标;②使用不合格项目排除清单;③要求项目进行生物多样性影响评估;④遵循缓解措施层级避免对当地生态系统造成损害,甚至在有可能的情况下,使当地生态系统受益;⑤纳入当地利益相关方。

这些方法不需要开发贷款方向借款国发号施令或一手包办项目。与此相反,这些方法以各国分享的经验和积累的专业知识为基础,能够推动国际合作以实现共同的目标。例如,国际机构开发了各类平台,用于协助相关方采用协作的方式实施缓解措施层级。世界自然保护联盟(IUCN)与世界资源研究所(WRI)联合开发的恢复机会评估方法(ROAM)就属于此类机制。ROAM 支持各国政府制订生态系统恢复计划,包括确定优先领域、对干预方法进行成本—效益分析,确定融资方案等[37]。在这方面,绿色"一带一路"提供了一个适宜的平台,可以促进国际合作,提升生物多样性保护在发展规划中的优先级,展示中国的经验,强化共建"一带一路"国家能力。

在上述五大方法中,中国的生态保护红线与遵循缓解措施层级的相似度最高。生态保护红线强调通过划定生态脆弱区和重要生态功能区避免对生态系统造成伤害。这些方法可以形成合力,也就意味着中国的生态保护红线可以纳入"一带一路"倡议。其他共建"一带一路"国家对缓解措施层级较为熟悉,而生态保护红线在中国国内已经取得了巨大成功。因此,在国际合作的基础上将规划生态保护红线的措施纳入"一带一路"项目规划可以有力地衔接"一带一路"倡议和 SDG15。联合国可持续发展解决方案网络执行主任吉多·施密特·特劳布(Guido Schmidt-Traub)[38]2020 年 3 月在《中外对话》上发表了《向中国学习"生态红线"》一文,文中指出,中国是目前唯一一个全方位且雄心勃勃地对土地用途进行规划的国家,对于任何希望实现生物多样性保护目标和气候目标的国家来说,中国的经验都值得借鉴。将土地利用规划纳入气候和生物多样性战略有助于《生物多样性公约》第十五次缔约方大会和《联合国气候变化框架公约》第二十六次缔约方大会取得成功,这种土地利用规划可成为新冠肺炎疫情背景下,引导刺激经济朝着正确方向发展的关键性工具。

3 SDG15 相关投资工具的分析

要实现 SDG15 并非易事。生物多样性对于人们的生活和生产来说必不可少，与此同时又十分脆弱，一旦被破坏就很难（或者说几乎不可能）再生。因此，为提高它在不断发展的国际金融投资界中的优先级，生物多样性融资应运而生。

人们的有关需求日益急迫。在 2015 年于北京举办的一个研讨会上，生物多样性和生态系统服务政府间科学政策平台（IPBES）得出结论，为了保护当前受土地退化影响的 32 亿人的利益，我们必须"协调一致，立即行动"，以遏制全球范围内生态系统的退化[39]。从经济角度来看，预计因生物多样性退化而导致的损失将占全球 GDP 的 10%。有研究人员追踪记录了全球各个国家和地区生态系统恢复的成功经验。

根据定义，保护生物多样性是一种将长期福祉置于短期繁荣之上的行为，它要求人们对支持未来经济生产和人类健康所必需的自然资本进行投资，同时，要求对那些能够在社区内产生广泛的正外向性的活动进行投资——投资者自身是无法完全消化这些正外向性的。因此，保护生物多样性需要靠外部激励来实现繁荣，如有利的政策环境、优惠的财政条款、有动力推动积极变革的有影响力的投资者等，这些有影响力的投资者不仅提高了自身的资产价值，也为他们所在的社区谋福祉。

3.1 中国经验的调查与评估

SDG15 旨在保护、恢复和促进可持续利用陆地生态系统。近年来，中国围绕健全生态补偿机制、生态功能区转移支付、草原奖补、退耕还林补贴、湿地保护修复补贴等不断加大财政资金投入，同时，不断健全自然资源产权制度，创新政府与企业、环保组织间的合作方式，推动森林可持续管理建设，防治荒漠化，制止和扭转土地退化，遏制生物多样性的丧失。2018 年，中国 SDG15 得分为 62.7，较 2017 年提高了 7%，陆地生态系统保护取得了一定的成效。

生态补偿机制不断完善[40]。中国政府高度重视生态补偿机制建设，《关于健全生态保护补偿机制的意见》《关于加快建立流域上下游横向生态保护补偿机制的指导意见》《建立市场化、多元化生态保护补偿机制行动计划》《关于建立健全长江经济带生态补偿与保护长效机制的指导意见》《生态综合补偿试点方案》等重要政策文件密集出台，具有中国特色的生态补偿制度格局日益清晰。据统计，2019年，中国生态补偿财政资金投入已近2 000亿元。同时，国家和地方都在积极探索市场化、多元化生态补偿机制，弥补政府财政补偿资金的不足。如南水北调中线水源区积极开展具有生态补偿性质的对口协作，浙江金华与磐安率先实践异地开发的补偿模式，新安江流域引入社会资本参与生态补偿项目，茅台集团自2014年起计划连续十年累计出资5亿元参与赤水河流域水环境补偿，三峡集团正在长江大保护中发挥主体平台作用并探索市场化补偿路径。

生态功能区转移支付力度不断加大。 为了引导地方政府加强生态环境保护力度，提高国家重点生态功能区所在地政府基本公共服务保障能力，自2008年开始，中央财政设立国家重点生态功能区转移支付，不断加大对重点生态功能区的保护力度，截至2019年累计下达转移支付资金5 242亿元。其中，2019年下达811亿元，较2018年增加90亿元，增幅达12.5%（图3-1）。同时，中国不断扩大国家重点生态功能区范围，2019年享受国家重点生态功能区转移支付县域数量已达819个。在纳入国家重点生态功能区后，地方政府将获得相关财政、投资等政策支持，必须严格执行产业准入负面清单制度，强化生态保护和修复，合理调控工业化城镇化开发内容和边界，保持并提高生态产品供给能力。

森林生态效益补偿标准不断提高。 近年来，中央财政不断加大森林生态效益补偿投入规模，逐步提高补偿标准。从2010年起，中央财政依据国家级公益林权属实行不同的补偿标准：国有的国家级公益林补偿标准为5元/（年·亩）[①]、集体和个人所有的国家级公益林补偿标准由原来的5元/（年·亩）提高到10元/（年·亩）。2013年，将集体和个人所有的国家级公益林补偿标准进一步提高到15元/（年·亩）。2015年、2016年、2017年将国有的国家级公益林补偿标准逐步提高到6元/（年·亩）、8元/（年·亩）、10元/（年·亩）。在中央财政不断加大投入、提高标准的同时，地方财政也应积极完善地方森林生态效益补偿制度。

---

① 1亩约等于0.067公顷。

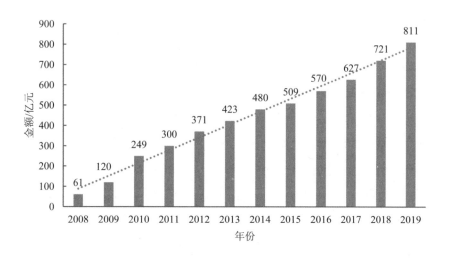

图 3-1　2008—2019 年国家重点生态功能区转移支付增长情况[41]

湿地生态保护修复财政扶持政策不断完善。中国高度重视湿地保护工作，不断加大投入力度，加快建立健全湿地保护修复财政扶持政策。2013—2016 年，中央财政共计投入 50 亿元支持中国湿地保护，后继续通过林业改革发展资金支持湿地保护恢复。2014 年，财政部和国家林业局启动湿地生态效益补偿试点，对候鸟迁飞路线上的重要湿地因鸟类等野生动物保护造成的损失给予补偿。目前，相关中央财政资金采取"切块"方式下达，由地方自主确定湿地生态效益补偿范围和湿地保护对象。

> **专栏 3-1　林业改革发展资金支持湿地保护修复的举措**
>
> 一是支持湿地保护与恢复。在林业系统管理的国际重要湿地、国家重要湿地以及生态区位重要的国家湿地公园、省级以上（含省级）湿地自然保护区，支持实施湿地保护与恢复，促进改善湿地生态状况，维护湿地生态系统的健康。
>
> 二是支持退耕还湿。支持林业系统管理的国际重要湿地、国家级湿地自然保护区、国家重要湿地范围内的省级自然保护区实施退耕还湿，扩大湿地面积，改善耕地周边生态状况。
>
> 三是支持湿地生态效益补偿。对候鸟迁飞路线上的林业系统管理的重要湿地因鸟类等野生动物保护造成的损失给予补偿，调动各方面保护湿地的积极性，维护湿地生态服务功能。

草原生态保护补助奖励政策持续推进。 为保护草原生态,保障牛、羊肉等特色畜产品供给,促进牧民增收,中国政府自 2011 年开始实施草原生态保护补助奖励机制。目前已覆盖内蒙古、新疆、西藏、青海、四川、甘肃、宁夏和云南 8 个主要草原牧区省(区),以及黑龙江等 5 个非主要牧区省,共计 268 个牧区半牧区(县),累计下达补贴资金 1 520.3 亿元。其中,2019 年,中央财政安排新一轮草原生态保护补助奖励 187.6 亿元(图 3-2),支持实施禁牧面积 12.06 亿亩、草畜平衡面积 26.05 亿亩。

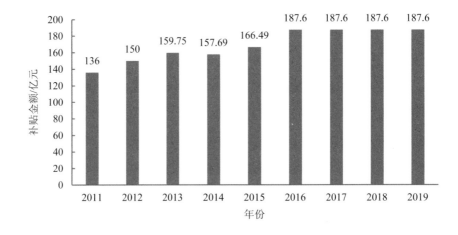

图 3-2　2011—2019 年草原生态保护补助奖励情况

资料来源:生态环境部环境规划院. 中国环境经济政策进展年度报告:2017,2018;财政部网站。

自然资源统一确权登记全面推进。 自然资源确权登记工作是推动自然资源资产产权制度改革的基础环节,而健全自然资源资产产权制度是中国生态文明制度建设的重要内容。截至 2018 年 10 月底,12 个省级行政区、32 个试点区域共划定自然资源登记单元 1 191 个,确权登记总面积 186 727 平方千米,并重点探索了国家公园、湿地、水流、探明储量矿产资源等确权登记试点。自 2018 年年底开始,在全国全面铺开、分阶段推进重点区域自然资源确权登记,计划利用 5 年时间完成对国家和各省(区、市)重点建设的国家公园、自然保护区、各类自然公园(风景名胜区、湿地公园、自然遗产、地质公园等)等自然保护地的自然资源统一确权登记,同时对大江大河大湖、重要湿地、国有重点林区、重要草原草甸等具有完整生态功能的全民所有单项自然资源开展统一确权登记。

创新环保组织和政府间、大型企业和政府间合作方式。"债务换自然"最早源于 20 世纪 80 年代，是非政府组织与各国政府共同开展的，将国家的债务款项转而用来作为该国自然保护活动资金的一种活动，其目的是实现发展和保护的共赢。目前中国国内尚未查到"债务换自然"的交易活动，但国际组织和政府间、大型企业和政府间以生态系统服务付费（PES）的形式开展了很多实现环保公益目的与投资者获取合理收益共赢的项目。以大自然保护协会（TNC）与浙江龙坞合作开展的水源地保护项目为例，该项目以 TNC 为公益顾问，通过信托机构出资进行水源地保护，并对林地进行统一经营管理，从中获取资金支付农户补偿金和水源地保护管理费用的方式，实现了水源地保护与信托公司合理收益的协同发展。

> **专栏 3-2　TNC 与浙江龙坞合作开展基于信托模式的水源地保护项目**[42-43]
>
> 　　2015 年 1 月 15 日，TNC 和浙江省黄湖镇政府签署了在龙坞水库开展村镇饮用水水源地保护的合作协议，降低龙坞水库水质可能受到的威胁，将水库水质从 Ⅱ 级提升到 Ⅰ 级，并思考实现生态保护与社区发展的双赢模式。该项目由阿里巴巴基金会资助，是 TNC 中国第一个基于信托模式的水源地保护项目。2015 年 9 月中旬万向信托—中国自然保护公益信托决定投入 33 万元支持该项目。
>
> 　　2015 年 11 月，万向信托推出中国首只水上基金信托——万向信托—善水基金 1 号。善水基金信托邀请跨国公司担任顾问。同月，信托龙湖小水源保护计划支持的首个水源保护管理项目正式启动。
>
> 　　善水基金信托创新性的一个重要体现在于对社会资源的高效整合，其将农户、金融机构、公益组织、当地社会团体、农业相关产业链下游企业以及消费者等多种角色共同纳入了其日常运营中，形成了互动、协作、共享的良好局面。该项目一方面解决了水库人为污染问题，使周边社区居民受益；另一方面创造了可持续的资金机制，同时实现环保公益目的与投资者获取合理收益的协同发展。
>
> 　　参与模式：信托模式；
>
> 　　补偿对象：附近社区居民；
>
> 　　补偿主体：阿里巴巴基金会；
>
> 　　补偿方式：农户将林地以信用托付方式进入信托，获得稳定的补偿金；
>
> 　　资金机制：信托机构对林地进行统一经营管理，同时推进生态竹笋种植和消费者生态体验，从中获取资金支付农户补偿金和水源地保护管理费用。

3.2 国际经验的汇总与评估

全球生物多样性融资有多种形式。第一种最常见的是生态系统服务付费（PES）和确保农业可持续性而进行的对土壤和水健康投资。PES 在发达国家和发展中国家都有应用，名称有所差异，如中国的"生态补偿"和欧盟的"农业环境项目"[44]。据《自然—可持续发展》2018 年刊登的一项全球调查估计，全球每年在 PES 上的投资超过 360 亿美元，其中大约 1/3 发生在中国[45]。360 亿美元大部分集中在流域补贴上，每年约 237 亿美元。在发展中国家中，厄瓜多尔的流域 PES 项目案例最为人熟知。2000 年，在大自然保护协会的帮助下，厄瓜多尔首都基多推出了全球首个市政水基金。基多的 PES 项目（以西班牙语"水保护基金"的首字母缩写命名为 FONDAG）建立了一个瓶装水工厂，对用户收取一定的附加费用，用于对为基多提供水资源的水域进行保护。

第二种常用方法"生物多样性抵偿"对绿色"一带一路"尤其重要。这些金融项目旨在通过为独立的生物多样性保护项目进行融资，减少新项目建设对生物多样性的净影响（或在可能的情况下对生物多样性产生净正影响）。亚洲、欧洲和美洲各国都出台了相关政策，支持此类生物多样性融资，但这些政策对相关项目的界定存在很大差异[46-48]。《自然—可持续发展》2018 年的一项研究只纳入了在净零损失政策下实施的项目，研究发现，在全球范围内共有将近 13 000 个此类项目，覆盖面积约为 153 679 平方千米。其中规模最大的几个项目位于不同的国家，包括蒙古国、巴西和乌兹别克斯坦[49]。但是，正如 2013 年 Gardner 等[50]在其《保护生物学》发表的文章中所述，净零损失标准在实践过程中难以实现，它要求以与丧失的生物多样性等量的新增生物多样性作为抵偿（不仅仅是保护），并且要求这些增量能够长期保持。要实现这些目标需要强大的机制支持和多地共同参与，以对冲掉部分项目失败所带来的风险。

从更广的范围上讲，"生物多样性抵偿"归属在 2.2.2"与 SDG15 相关的国际标准"中提到的缓解措施层级的框架下。作为"补偿"阶段的一部分，抵偿是最后的选择。只有当项目无法避免或最小化对生态系统和依赖这些生态系统的社区的影响，或无法修复（恢复）生态系统时，才会选择抵偿。例如，2014 年 Villarroyo、Barros 和 Kiesecker 对拉丁美洲关于生物多样性抵偿的国家政策进行了分析,并发现三个国家(智利、哥伦比亚和墨西哥)的政策规定在环境影响评估流程中同时特别提到了缓解措施层级和抵偿。但是,很多学者提到,那些寻求支持抵偿计划的政府还需大力加强机制和能力建设,尤其是在科学的基础上实现不同地理区域之间的"生态平衡",以确保抵偿计划实现净零生物多样性影响[49-52]。

然而,2009 年 Luck、Chan 和 Fay[53]发现,全球范围内的生物多样性融资面临着严重的地域调配失衡问题:大部分资金都流向了低优先级的生态系统,而那些最重要的生态系统却被忽略了。超过一半的资金流向了美国、加拿大和欧洲,但实际上对于保护生态系统服务和生物多样性这两个目标而言,这些生态系统的优先级相对较低。同时,对于实现上述两个目标来说优先级较高的地区大都集中在东南亚和南美洲,而它们获得的生物多样性融资总共不超过全球的 15%。出现这种失调与一个事实不无相关:在生物多样性融资中,有超过一半来自国内政府拨款,因此大部分资金都留存在富裕国家。所以,如果全世界要想在实现 SDG15 方面取得进展,通过投资与援助(尤其是援助)开展国际生物多样性合作至关重要。

与商业投资人的合作

生物多样性融资通常仅限于援助,包括官方发展援助(ODA)和慈善援助。然而,近年来,商业投资方也逐渐获得机会参与生物多样性融资。在中长期内,许多旨在维持或增强生物多样性的活动都能实现自给自足,但前期需要融资才能启动。从长期来看,通过保护或增强现有的自然资本,这些投资可以降低经济生产的成本。例如,2018 年 Burian 等[54]研究主张针对改善土壤健康、提高土地适应性来进行农业投资,这可以提高农作物产量,减少农用化学品投入的成本,进而增加经济效益。根据 IPBES 预测,保护土壤所带来的经济效益比保护投入的成本平均高 10 倍[39]。最终,随着水污染越来越轻微,城镇和农村人口的生活改善,积极影响将逐渐辐射到下游生态系统,保护活动产生的惠益也将成倍增加。

生物多样性融资若想获得成功,须采取符合当地需求的措施,根据当地投入进行设计,并由当地政府进行妥善管理。2018 年 Clark、Reed 和 Sunderland[55]发现,当前生物多样性融资领域受到潜在的"漂绿"影响,即商业投资者的活动并非为了加强或保护生物多样性,而是为了推销自己,以获得有利的融资并提升口碑。虽然少数此类行为可能不会造成太大危害,但若允许他们借"生物多样性融资"之名肆意发展,可能给整个领域带来风险:不仅会威胁其债权的合法性,还会使其失去获得有利融资的机会,无法持续发展。

综合考虑潜在的惠益和风险,联合国开发计划署(UNDP)"生物多样性融资倡议"(BIOFIN)确定了五个重点领域,用于构建适用于商业性生物多样性融资的体系框架。

- 政策和机构审查,检验国家机构是否有足够的能力和健全的体制来促进生物多样性融资,以及哪些领域能从改革中受益(增加了一个可选项,即找出导致生物多样性丧失的经济动因);
- 支出审查,统计当前用于支持生物多样性的支出;
- 需求评估,估算用于支持生物多样性的总需求额,及其与实际支出的差额;
- 融资计划,设定目标并寻找资金的可能来源;
- 融资解决方案,制订并实施计划以解决先前步骤中发现的体制及财务缺口。

与共建"一带一路"国家开展双边"融资多样性倡议"合作

随着生物多样性融资领域的不断发展,尤其是在对商业活动开放之后,中国有机会成为该领域的全球领导者。鉴于"一带一路"倡议的本质是全球性合作网络,保护沿线"热点地区"的生物多样性,确保"一带一路"为所有社区及相关生态系统带来净收益,至关重要。

在中国保护性融资的潜在合作伙伴中,有两个"热点国家"尤为引人注目,即位于太平洋两岸的厄瓜多尔和印度尼西亚。这两个国家都与中国签署了"一带一路"倡议谅解备忘录,两国在 17 个全球生物多样性特丰国家榜上有名,共拥有全世界 17%的物种,分别是全球陆地生物多样性和海洋生物多样性最丰富的国家。素有"地球之肺"之称的亚马孙雨林中生物多样性最丰富的地区就位于厄瓜多尔[56]。厄瓜多尔境内的亚马孙雨林位于亚马孙河的源头,因此,保护这里的生态系统也会使下游的生态群落受益。印度尼西亚的海洋生物多样性位列全球之首,这里的珊瑚种类繁多且密度很高,因此也经常被称为"海中亚马孙"或"珊瑚金三角"。就联合国开发计划署提出的 BIOFIN 而言,厄瓜多尔和印度尼西亚均已取得重大进展，已准备好接受和管理生物多样性融资。

此外，印度尼西亚和厄瓜多尔都与中国有着紧密的经济联系。根据 FDI Markets 的数据，过去十年间，中国在印度尼西亚的新增投资逾 520 亿美元，超过了在其他任何国家的投资。尽管厄瓜多尔比中国小得多，但在过去的十年间，中国一直是厄瓜多尔最重要的债权国，两国政府也结下了深厚的情谊。2019 年，厄瓜多尔正式加入亚投行，成为亚投行第一个拉丁美洲或加勒比海国家正式成员，表明了厄瓜多尔加强与亚洲，特别是与中国的金融联系的意愿和机构就绪度。

保护生物多样性所面临的一个主要障碍，可以概括为一个简单的地理问题：生物多样性丰富的热点地区大都位于发展中国家，而这些国家的财政空间往往非常有限，无法设计和实施需要多年才能产生经济效益的长期项目。要解决这一问题，方法之一就是让这些国家与其最重要的战略伙伴（债权国或投资国）开展双边或多边合作，确保其经济开发活动不会造成环境退化。在双边生物多样性融资领域，主要有三种模式：债务换自然、国家环境基金（NEF）和双边可持续发展银行。

在债务换自然模式中，债权国取消一部分债务，作为交换，债务国用于偿还债务的款项将被用在保护生物多样性上。另外，影响力投资者或国际非营利组织也可通过谈判交易发挥关键性作用：它们可以折扣价购买一国的债务，再与债务国合作建立制度性基础设施以监督生物多样性计划，并帮助其设立用于支持上述活动的基金。这些交易能够有效抑制"财政受限—环境管理不到位—经济效益受损—财政进一步受限"的恶性循环。

如果实施得当，债务换自然能够让长期负债的国家通过减少破坏环境的活动来偿还债务。他们还可以建立有关的体制结构，界定什么样的活动可被定义为适合新保护区的"可持续经济活动"；监督财政空间以确保新保护措施得以妥善管理；同时，当地社区也应充分参与进来，确保整个过程执行到位。但是，债务换自然并不是解决严重债务问题的特效药，也不能让正在发生的生态灾难立即停止。正如塞舌尔的案例所示，建立保护区需要多年的努力。因此，最好把它当作一个长期的、主动的保护方案，而不要将其视作灾难发生时最后的"救命稻草"。

国家环境基金模式与债务换自然模式有许多相同特征，但其外部合作伙伴的干预较少。国家环境基金是与外部合作伙伴协作建立的由本土管理的基金，用于支持国内的保护工作。国家环境基金具有"信托基金"性质，这使其特别适宜作为中长期投资项目的融资工具，如可用于支持国家公园的划定、建立与维护。例如，巴西的亚马孙基金会（Amazon Fund）支持了居住在森林里的社区实施不会造成毁林的生计项目[57]。其他亚马孙流域国家（包括玻利维亚、哥伦比亚和秘鲁）也建立了国家环境基金，用于支持其国内的保护地体系。在亚洲，不丹和菲律宾都建立了相似的基金[58]。

国家环境基金由中央政府管理，可以与海外的战略合作伙伴共同建立。例如，菲律宾环境基金会就获得了美国和日本通过债务互换给予的支持。在这些案例中，国家环境基金与债务换自然相类似，只是在具体条件上程度不同。债务国不需要承诺预留特定的土地进行保护，只需要对国内制定的保护战略提供广泛支持。地方政府负责监管资金，因此，国家环境基金更适宜支持债务国与那些希望地方政府尽可能发挥作用的合作伙伴开展双边合作。

最后，可以通过建立特殊用途的开发银行实行双边保护融资。例如，北美开发银行是美国政府与墨西哥政府共同实施的项目，是北美自由贸易协议（NAFTA）谈判成果之一，目的是确保自由贸易协定通过后，美墨边境不会因为相关活动增加而发生环境退化。北美开发银行为边界两侧开展的可持续发展项目提供资金[59]。截至 2018 年年底，已为项目融资 12 亿美元[60]。在建设跨境交通走廊（如"一带一路"倡议中的交通走廊）或国家之间希望大幅度提高投资与贸易量时，这种模式具有独特的吸引力。

> **专栏 3-3 塞舌尔群岛的"债务换自然"**
>
> 大自然保护协会的生物多样性融资平台 Nature Vest 成立于 2014 年，旨在鼓励私营领域资本参与生物多样性保护。2016 年，Nature Vest 联合其他私人投资者一起，与塞舌尔的巴黎俱乐部债权人签署了一项协议，以一定折扣购买了塞舌尔的部分债务——花费约 2 200 万美元买入了约 2 500 万美元的债务。
>
> 通过与塞舌尔政府的合作，这项债务减免将对约 40 万平方千米的海洋进行经营和维护。在撰写本专栏时，大约一半的区域已经被划出，形成了两个新的保护区，而剩下的部分预计也将于此后一年内加入规划。

> 有两个因素共同促成了此次"债务换自然"的成功，一是塞舌尔政府的领导，二是整个规划过程的不疾不徐。由于这两点原因，塞舌尔项目获得了当地的支持，这有利于之后几年内管理和实施工作的展开。
>
> 这一项目标志着塞舌尔政府于 2012 年发布的一系列国家目标达到最高成就。2012 年，政府宣布计划进一步扩大保护区范围，将其海洋专属经济区的 30% 包括在内。塞舌尔采用了符合国际最佳实践的绘图法，并根据联合国教科文组织（UNESCO）的建议进行了调整[61]：即开始于 2014 年的塞舌尔海洋空间规划（MSP）计划（"塞舌尔海洋"）。塞舌尔海洋空间规划的目的在于拉长时间线，确保最终的结果有充足的事实依据和公众参与。实际上，虽然第一阶段已于 2018 年完成，将 15% 的专属经济区（EEZ）划入了保护区，第二阶段预计要到 2020 年年底才能完成（"计划"，2019）。为明确要保护哪些海洋区域、允许哪些可持续活动，MSP 举办了 9 次公共研讨会和 60 次磋商会，共听取了 10 个部门和 100 位公共利益攸关者的意见。

3.3 在国际层面利用中国实践经验的适用性

在过去十年间，中国及全球范围内的生物多样性融资都出现了大幅增长。生态系统服务付费这一方法已在国际上广泛应用了很多年，中国主要在过去十年快速发展。其他国家也对其他类型的生物多样性融资进行了深入探索，如生物多样性抵偿和债务换自然。在这些领域，中国的专业知识可以帮助共建"一带一路"国家应对一些项目中遇到的挑战，中国也可以从成功的国际案例中吸取经验。

尽管很多国家开展了 PES 项目，但在深度、广度和发展速度方面，中国开发国内项目的经验具有较好的参考和借鉴意义。中国推出的造林计划（针对生态林和商业林的项目有区别）、退耕还草计划（针对畜牧区和混合区的计划有区别）以及湿地计划（包括对重要鸟类迁徙路线沿途土地给予特殊补贴）都具备高度的科学性和专业性，并得到了当地社区的理解。很多共建"一带一路"国家都可以根据自身实际情况加以借鉴。"一带一路"强调合作，这就意味着各国可以共享信息以及协同规划。在这样的背景下，尽可能分享相关经验是明智之举。

在全球范围内，生物多样性融资大多结合了行政和市场机制。在一些情况下，政府对于项目进行全权管理，尤其是针对那些关注环境服务付费的项目。但是在一些抵偿项目中,生物多样性增强与商业投资导致的生物多样性丧失相关联,政府通常关注为生物多样性抵偿机制建立法律框架。尽管生物多样性抵偿在发达国家和发展中国家都越来越受欢迎,但同时对多个地理区域(有些区域的生物多样性丧失,有些地区增强)就需要强大的机构能力。在这一过程中需要掌握足够的科学技术和专业知识来进行生态调查,以准确把握不同区域之间的平衡,并对结果进行评估。尽管中国并没有通过抵偿将生物多样性增强项目和生物多样性丧失建设项目联系在一起,但中国同样需要加强这方面的机构能力建设,以建立和完善生态保护红线体系。正在探索生物多样性抵偿机制的共建"一带一路"国家或许可以在该领域寻求合作机会。

中国国家开发银行与联合国开发计划署 2019 年联合完成的一项研究显示,在共建"一带一路"国家和地区开展合作协调各类标准,可以产生诸多惠益,详见表 3-1。本研究也发现,通过协调标准,分享中国生态保护红线体系建设技术、合作伙伴生物多样性融资经验,共同确定保护的优先领域,所有合作方都能够获益。

表 3-1　协调共建"一带一路"国家和地区标准可以产生的惠益

国家	受益方	惠益
中国	政府	• 预防并减少对声誉造成的风险; • 预防并减少环境与社会风险
	金融机构	• 防范资金回收风险; • 提供更具创新性和竞争力的金融服务
	企业	• 市场得到扩展; • 项目回报更高、效果更好; • 竞争力增加,风险管理状况得到改善
合作方	政府	• 经济发展,贫困减少; • 财政资源管理得到改善; • 自然资源管理得到改善; • 机构能力得到强化; • 预防并减少环境与社会风险; • 遵约成本减少

国家	受益方	惠益
合作方	企业	• 扩展国内市场，参与国际价值链； • 参与"一带一路"相关招标与采购项目的机会增加； • 遵约与风险管理状况得到改善
全体	当地社区	• 生计改善，体面工作； • 发生社会冲突的可能性减小； • 对项目的主人翁意识提升，发声增强； • 面对潜在负面影响冲击时的脆弱性减少
	全球社区	• 资源利用与发展更加公平； • 互联互通与合作增加； • 提供全球公共产品； • 全球治理状况得到改善

资料来源：根据联合国开发计划署及中国国家开发银行（2019）研究总结。

4 SDG15 相关治理结构的分析

4.1 中国经验的调查与评估

4.1.1 中国生物多样性保护治理架构

中国将生物多样性保护作为生态文明建设目标体系的重要内容,生物多样性保护体制机制不断完善。中国生物多样性保护实行国家统一监管和部门分工负责相结合的机制,特别是从 1993 年中国批准《生物多样性公约》以来,成立了由原国家环境保护总局牵头,有国务院 20 个部门参加的中国履行《生物多样性公约》工作协调组,在原国家环保总局成立履约办公室,建立了国家履约联络点、国家履约信息交换所联络点和国家生物安全联络点,并建立生物物种资源保护部际联席会议制度。履约工作协调组每年都召开会议,制订年度履约工作计划,开展了一系列形式多样的活动,初步形成了生物多样性保护和履约国家工作机制。中国于 1992 年开始编制《中国生物多样性保护行动计划》(以下简称《行动计划》),并于 1994 年正式发布。《行动计划》确定了中国生物多样性优先保护的生态系统地点和优先保护的物种名录,并明确了七个领域的目标。

2010 年,中国国务院成立了"2010 国际生物多样性年中国国家委员会",召开会议审议通过了《国际生物多样性年中国行动方案》和《中国生物多样性保护战略与行动计划(2011—2030 年)》。2011 年 6 月,国务院决定把"2010 国际生物多样性年中国国家委员会"更名为"中国生物多样性保护国家委员会",国务院分管副总理任主任,目前成员单位共 23 个,统筹协调全国生物多样性保护工作,指导"联合国生物多样性十年中国行动"。中国生物多样性保护国家委员会的成立表明了中国加强环境保护、推进可持续发展的决心,也是对国际社会的庄严承诺。自 2015 年以来,中国已出台或修订与生物多样性保护相关的政策法规 56 部,生物多样性保护政策与法律法规体系日臻完善。

除中央政府的治理架构之外,中国省级地方政府环境保护机构改革也呈现有利于进一步保护生物多样性的趋势。2008 年,国家环境保护总局升级为环境保护部,成为国务院组成部门,各省、自治区和直辖市将环境保护机构提升为环境保护厅,体现了环境保护系统管理体制的统一性。一些省级政府也相继建立了生物多样性保护的协调机制。仿照国家层面对生物多样性管理职责定位,规定环境保护厅行使本辖区"牵头"管理生物多样性的职责,并将管理职责落实到了环境保护厅内设机构。为适应本地生物多样性统一管理的需要,有些地方设立了符合本地生物多样性实际的本土管理机构,如云南省设立了湖泊保护与治理处,体现了本地机构改革和生物多样性管理特色。2018 年,根据《中共中央关于深化党和国家机构改革的决定》,中国国务院新组建了生态环境部,体现了统筹山水林田湖草沙系统治理的整体观。各省、自治区和直辖市新组建生态环境厅,全面指导、协调和监督生态环保工作。

中国生物多样性保护的政策执行、治理框架要素以及生物多样性保护的其他重要机构详见附录 4。其他重要机构主要包括中国生物多样性保护国家委员会和中国科学院生物多样性委员会等。

4.1.2 绿色"一带一路"与生物多样性保护

基于中国生物多样性保护经验与绿色"一带一路"建设的需求,已有相关措施推动两者在治理机制、治理体系、信息、技术科研合作、绿色投资与金融等方面的进一步关联与协同,促进共建"一带一路"国家的生物多样性保护和 SDG15 的落实。

第一,建立合作治理机制和平台,推动完善共建国家生物多样性治理体系。以绿色"一带一路"建设为统领,统筹并充分发挥现有双边、多边国际合作机制,构建生物多样性保护网络,创新合作模式,建设以各国政府、智库、企业、社会组织和公众共同参与的多元合作平台,强化中国—东盟、上海合作组织(SCO)、澜沧江—湄公河合作、亚信非政府论坛、欧亚经济论坛、中非合作论坛、中国—阿拉伯合作论坛等合作机制作用,推动六大经济走廊的环保合作平台建设,扩大与相关国际组织和机构合作,促进 SDG15 的有效落实。

第二，促进绿色技术、科研合作。加强绿色、先进、适用技术在"一带一路"沿线发展中国家转移转化，促进先进生物多样性保护技术的联合研发、推广和应用。打造科研机构、智库之间的科学研究和技术研发平台。与相关国家和地区展开生物多样性联合研究将为全球生物多样性保护提供契机。通过对沿线国家和地区进行生物多样性科学考察，分析区域内生物多样性进化机制和地理分布特征与模式，将进一步促进全球生物多样性的科学认识，为"一带一路"沿线国家的青年官员、科学家提供培训与能力建设。

第三，推动信息交流。生物多样性保护信息共享和公开，提供综合信息支撑与保障。加强绿色"一带一路"生态环保大数据服务平台中有关生物多样性信息库的建设，发挥国家空间和信息基础设施作用，推动环保法律法规、政策标准与实践经验交流与分享，加强各国部门间统筹合作与项目生态环保信息共享与公开，提升对境外项目生态环境风险评估与防范的咨询服务能力，推动生态环保信息产品、技术和服务合作，为绿色"一带一路"建设提供综合环保信息支持与保障。

第四，促进绿色投资、绿色贸易和绿色金融体系发展。绿色金融体系的建设有助于为"一带一路"项目的长期运行打下良好基础。以中国—东盟投资合作基金发布的《关于在东盟地区投资的社会责任与环境保护参考指引》为例，规定东盟投资合作基金（CAF）根据自身环境和社会管理系统（ESMS），建议企业在对外投资时，可参照"绩效标准"来识别、管理环境和社会风险的影响，明确投资过程中的评估指标，及投资后期的持续监督，推动被投资公司以可持续的经营方式避免、缓解、管理风险。这一"绩效标准"包括生物多样性保护和生物自然资源的可持续管理等共8个方面，共同确定了客户在整个对外投资的项目周期内需达到相关生物多样性可持续管理等标准，具体包括：①确认公司是否就项目涉及的生物多样性的影响进行了解和处理。②确认公司是否在受法律保护的地区进行活动。③确认在项目执行过程中是否会引入外来物种。如果有引入外来物种的计划，应确认是否已经收到相关政府监管部门的批准。④确认项目所需要利用到的自然资源、森林及植被、淡水和海洋资源是否为可再生资源，并致力于以可持续的方式管理它们。

第五，《"一带一路"绿色投资原则》（GIP）等强化了对企业行为的绿色指引，鼓励企业采取自愿性措施保护环境，推动可持续发展；鼓励环保企业开拓沿线国家市场，引导优势环保产业集群式"走出去"，可借鉴中国的国家生态工业示范园区建设经验与标准，加强生物多样性保护，优先采取就地、就近保护措施，做好生态恢复；引导企业加大应对气候变化领域重大技术的研发和应用。

第六，推动"一带一路"合作中的性别平等，强化生物多样性保护中的女性领导力。生物多样性与性别在国际上属于热点前沿问题。近年来，在生物多样性保护工作中推动性别主流化得到了国际社会的广泛关注，"生物多样性与性别"已经作为热点议题写入《生物多样性公约》。但目前中国在生物多样性研究领域存在机制不完善、意识较弱等问题。建议全面提升相关机构的能力建设，在各相关部门设置性别联络人，建立跨部门性别主流化交流合作机制；在生物多样性管理部门和机构开展性别主流化培训，提升工作人员的基本意识；在生态环境保护、绿色"一带一路"建设相关政策中考虑性别因素，在具体项目中设置对性别因素的考核指标。这些做法也将帮助"一带一路"合作项目满足投资中的性别相关国际标准和东道国相关要求，促进民心相通，助力"一带一路"建设行稳致远。

4.2 关于生物多样性保护治理的国际经验汇总与评估

2.2节指出，国际环境管理体系在过去几十年中迅速发展起来，并列举了国际金融机构的分析与评估体系，本节将重点分析所涉国际金融机构的问责机制。

全球各地的金融机构已经行动起来，共同为实现SDG15而努力，同时确保机构活动能够保护受到项目影响的生物多样性。2.2节阐述了与SDG15相关的标准和指导方针，而本节将阐释金融机构为充分考虑生物多样性保护需求而采用的治理结构。本节对中国政策性银行的同业机构所采用的治理结构进行了比较，即大型多边金融机构，具体包括AFDB、ADB、AIIB、EBRD、EIB、IDB、IFC、KFW和WB。

4.2.1 生物多样性治理：将SDG15纳入金融机构的决策机制

如2.2节所述，大多数主要的金融机构都是通过使用既定标准、遵循缓解措施层级、与受影响的利益相关方磋商等方式，将生物多样性的有关因素纳入其运营过程。这些利益相关方很可能依赖当地的生态系统谋生，因此对任何的生物多样性威胁都十分敏感。除此之外，一些金融机构还在SDG15的基础之上增加了其他步骤。不同的机构之间做法也存在很大差异。但是，各机构在这方面为自己设定的要求出现了越来越多的共性，包括：

- 将评估与专业性结合：非洲发展银行和亚投行要求资深专家提供意见，找出那些可能受到影响的生态系统和生态系统服务功能。
- 提升项目实施者能力以应对不断变化的情况：非洲开发银行、亚投行、欧洲投资银行和世界银行都要求在其项目中使用"适应性管理"。在这种情形下，借款人和客户必须考虑在项目开展过程中，外部条件可能与最初的预期不一致这种可能性。可能会发现新的物种或出现其他与生物多样性相关的项目影响。在项目规划中，应具体说明可能出现哪些类型的挑战，以及项目实施者将如何适应变化的情况。规划完成后，实施人员有权在项目过程中对计划进行调整。以亚投行为例，重大变更需要进行额外的环境评估，以确保计划调整得当。

4.2.2 政策实施：监测和报告

借款人和客户可能承诺将以负责的态度进行环境管理，金融机构则努力把对生物多样性的影响考虑在内，但决定最终结果的是实际表现。为此，金融机构通常会为其借款人和客户制定有关的监测和报告要求。这样做的同时，金融机构也会强调对借款国国家主权的尊重，设计出能让贷方和借款方之间合作达到最佳结果的可行方案。金融机构有几种不同的做法，包括让借款人承担主要的监测责任、借助外部审计等。

4.2.3 政策实施：申诉机制

许多金融机构（包括多边机构和国家性机构）都有相关政策，如果利益相关方（包括独立的非政府组织和项目受益方）质疑金融机构支持的项目在实施过程中破坏了当地的生物多样性，他们可以提出申诉，并要求进行调查。通过建立相关体制机制进行听证、调查和裁决。金融机构可以确保其借款人和受赠人遵守协议规定，防止小范围破坏演变为大范围破坏，维护自身在全球的声誉，并从中吸取经验教训以供未来项目借鉴。

这些申诉机制可能是机构级别或项目级别的，也可能二者兼有。项目级的申诉机制灵活度更高，对于使用集中式系统解决来自世界各地项目申诉的做法而言，其速度更快，且对利益相关方来说更易操作。但对于金融机构而言，项目级的申诉机制管理起来可能更为繁琐，因为需要同时监督许多不同国家的进程。

附录 4 中的表 A4-1 描述的是项目级申诉机制设计中的常见元素：机构定位、资源、设计与建设、流程和对申诉者的保障等。

在机构级的申诉机制中，利益相关方可以将申诉提交至金融机构的主体或其指定的投诉机构。对于金融机构来说，这类申诉机制更易管理，因为仅涉及一个机构的创建和管理。但是对于受到影响的利益相关方而言，这类服务不仅较难获得，而且耗费的时间往往会比项目级的申诉机制更长。

表 A4-1 显示了金融机构在项目申诉机制方面考虑的不同政策因素。这些制度安排的巨大差异使得金融机构在设计自己的机制时可以相互借鉴。

表 A4-1 中列出的所有金融机构都设有机构级的申诉机制，但由于形式太过多样，无法一一以表格形式呈现。除这些金融机构外，其他几个主要的多边和国家开发银行也有这类机制，包括 IDB、CAF 和 BNDES。

4.2.4　纳入性别视角

无论用于哪种场合，在设计问责机制时必须要考虑妇女因素，国际金融机构非常了解这一点的重要性。在许多农村贫困地区，妇女的财产所有权有限，她们的财产所有权通常是经由其父亲、丈夫或儿子进行登记的。在这种情况下，因为她们可能无法证明其财产价值的损失，国家司法系统难以承认妇女有资格通过地方法院提起诉讼。但如果不听取她们的诉求，基于性别的生物多样性风险可能会被忽视和恶化。亚洲开发银行与世界银行均建议其项目要确保妇女均可使用问责机制，而不考虑财产所有权问题[17,62]。这就完成了将性别因素纳入生物多样性筹资的全过程，以确保相关项目不会对妇女造成差异化影响，以及削减妇女在地方开展生物多样性管理的能力。表 4-1 收集了国际金融机构在整个项目周期中纳入性别因素的最佳实践。表中所列内容不一定完整，仅展示了国际金融机构的研究与评估人员记录在案的常用最佳实践。

表4-1 将性别视角纳入生物多样性融资的最佳实践

项目阶段	最佳实践
上游阶段：规划	在评估当地生物多样性的预期损失以及社区利用当地生态系统的方式变化时，分性别评估对当地生计的预期影响。确保妇女在开展传统的采集工作时不会遭遇更大的困难。在男女承担不同传统工作任务的情况下，这种做法格外有效。在设计利益相关方参与程序时，确保妇女可以充分参与。这种做法可以帮助规划人员理解项目可能对男女产生差异化影响的不同方式。在女性通常不参加公开讨论的环境中，需考虑规划仅供女性参与的空间
中游阶段：实施	如项目需对社区因无法再利用当地生态系统而进行金钱补偿时，应确保合理分配经济补偿，不损害妇女福祉。在妇女通常支配其从当地生态系统中采集的资源而男子控制钱财的环境中，这种做法尤其重要
下游阶段：监测与问责	考虑男女在利用时间与钱财方面的变化。妇女通过在家庭菜园或乡村菜园中种植传统农作物品种来维持农业生物多样性，在这种情况下，这种做法可以确保生物多样性不受影响。在发生极端天气事件或经济动荡时，菜园的农作物生物多样性是当地粮食系统恢复力的关键。 确保妇女可以完全使用问责机制和申诉机制。在妇女缺乏平等财产权、使用地方司法系统的机会有限或通常不参加公共讨论的情况下，这种做法尤其重要。妇女参与问责机制可以使项目监管人员和出资方监测项目对妇女维持农业生物多样性这一传统工作的影响。 在项目结束后进行评估时，制定一个"提示表"，以便在未来的项目规划中，可以在此类特定情况下纳入性别因素。以此方式不断积累经验，有助于确保未来在同样的文化背景下实施发展项目时，能够充分吸收这个项目提供的经验

4.3 中国与国际的异同

无论是在中国还是在国际上，与SDG15相关的治理结构都包括行政和执法机制。在中国国内，行政措施的制定和实施需要各部委和其他政府部门之间的合作，司法系统是政策实施和执行的主阵地。在国际上，其他国家以类似的方式合作，对国际参与者的标准和期待进行协调。国际执法和申诉机制与各国国内法院的功能类似，在保证各方参与的同时优先调解纠纷。

在行政方面，中国的生物多样性保护相关工作由生态环境部牵头。此外，中国国务院还成立了由生态环境部等23个部门组成的"中国生物多样性保护国家委员会",从国家层面对生物多样性相关行动进行监督,将生物多样性纳入了社会经济发展和各部门的日常管理,极大地促进了生物多样性在中国的主流化。此外,各省、自治区和直辖市也将环保厅/局重新组建为生态环境厅/局,形成了一个全国统一强化的网络。

司法系统是中国政策实施和执行的主阵地。最高人民法院设立了专门的环境资源审判庭,对下级人民法院环境资源民事案件审判工作进行指导,研究起草有关司法解释等。此举有助于统一完善全国范围内的环境纠纷调解。

国际上,多边政府间组织和金融机构已通过多种方式将生物多样性治理纳入项目管理中:采纳项目利益相关方和独立专家的反馈意见;提升项目经理能力,助其适应不断变化的条件;建立问责和处理申诉的体制机制;将性别因素放在整个项目周期中考虑。行政机制主要采用项目上游阶段所用的规划手段,将强制标准、缓解措施层级和利益相关方参与措施纳入在内。由于中国的国际合作伙伴包括多边机构和国家机构、政府和金融机构,"一带一路"倡议采用了协调环境与社会风险管理的方法,从项目启动就确保协调各方期待,从而在整个项目周期维持良好的合作关系,保护生态系统。随着"一带一路"的涉及面不断扩大,协调各方期待的重要性也将日益凸显。"一带一路"倡议强调协作共建,这就意味着中国相关部委可以与"一带一路"参与国家政府部门代表合作,积极参与标准的制定。

与"一带一路"项目主要的中国贷款方(国家开发银行和中国进出口银行)相比,国际金融机构建立了一系列监测和申诉机制,以确保行政措施的有效性。这些措施是协作性质的,可以作为当地社区、项目所在国政府和项目执行承包商解决纠纷的平台。与国内法庭类似,它们能够加强民众对相关治理机构的信任。设计和纳入类似的环境纠纷解决机制对"一带一路"的发展大有裨益。

5 政策建议：绿色"一带一路"建设路线图

在前面章节中，研究团队梳理了中国和其他国家在平衡投资项目惠益与项目带给社区及生态系统的风险方面取得的进展。考虑"一带一路"倡议的发展速度和覆盖范围，至关重要的是要利用所有可能的经验来实现额外的增长，以确保"一带一路"倡议发挥其潜力，助力全球可持续发展。在第一期绿色"一带一路"专题研究成果的基础上，本章进一步完善了绿色"一带一路"建设的路线图，并提出了对接"一带一路"与 SDG15 的政策建议。

5.1 绿色"一带一路"建设路线图

5.1.1 加强政策沟通，将绿色"一带一路"建设作为落实 2030 年可持续发展目标和推进全球环境治理变革的重要实践

始终坚持把绿色作为"一带一路"建设的底色。坚持将绿色发展理念和生态文明思想贯穿到"一带一路"建设"五通"的方方面面，推动绿色基础设施建设、绿色投资、绿色金融，将"一带一路"建设成为绿色与可持续发展之路，构建以绿色发展为基础的人类命运共同体。

推进"一带一路"国际多边合作平台绿色化。在"一带一路"国际合作高峰论坛中常态化设置"绿色之路"分论坛，发挥"一带一路"绿色发展国际联盟和"一带一路"可持续城市联盟作用，搭建共建"绿色丝绸之路"的国际合作平台，助力全球层面落实 2030 年可持续发展目标和完善全球环境治理体系变革。在共建"一带一路"国家推广绿色发展理念与实践，开展国家、城市和项目绿色发展试点示范。此外，发挥"一带一路""五通"功能，共同推动生态环境保护与应对气候变化等相关政策的落实，支持《生物多样性公约》《濒危野生动植物物种国际贸易公约》《联合国气候变化框架公约》等环境国际公约进程。

5.1.2 加强战略对接,建立绿色"一带一路"与联合国可持续发展议程战略对接机制

鉴于绿色"一带一路"倡议是推动落实联合国可持续发展议程,特别是开展国际生物多样性保护的重要工具,建议通过以下步骤实现相关规划与生物多样性目标的战略对接。

加强顶层设计。将推进落实联合国 2030 年可持续发展目标作为绿色"一带一路"建设的重要任务,在与有关国家和国际组织签署合作共建"一带一路"谅解备忘录时,把共建绿色"一带一路"、促进"一带一路"建设与联合国 2030 年可持续发展目标对接作为重要内容。

建立推进机制。根据不同国家实际,与合作方设立工作组/专家组,共同拟订共建"绿色丝绸之路"的发展战略,结合相关国家落实 SDGs 的实际需求,确定近中远期合作重点领域并做好相关规划的衔接。

建立参与和反馈机制。构建政府引导、企业支持和社会参与的支持网络,重点完善相关国际组织参与机制,建立包括协商、决策参与和动态反馈在内的全过程参与机制,确保绿色"一带一路"与 2030 年可持续发展议程在开放透明的环境下顺利对接。

建立沿线城市和地方的专业化合作机制。鼓励沿线城市根据各自产业结构特色和发展目标定位,就共性问题制定支持政策框架,发掘合作机会,引导企业参与共建合作。

5.1.3 加强项目管理,建立完善绿色"一带一路"项目管理机制

为将上述战略纳入"一带一路"项目管理,建议采取以下步骤。

建立"一带一路"项目生态环境管理机制。加强中国与共建国家之间、中国政府主管部门之间的沟通协调,建立以科学为基础的应对各类风险的项目风险评估和管理机制,在项目设计、建设、运营、采购、招投标等环节严格遵守东道国标准,并鼓励项目采用国际组织、多边金融机构实施的有关生态环境保护的原则、标准和惯例,努力实现高标准、惠民生、可持续的目标。支持金融机构将生态环境影响因素重点纳入项目评级体系和风险评级体系,建立"一带一路"项目环境

与社会风险评估方法和工具,作为政府管理部门、开发性金融和政策性金融支持的重要标准,鼓励商业金融参照执行。

倡导"一带一路"框架下广泛采用绿色金融工具。一是探索建立"一带一路"绿色发展基金,重点支持沿线国家生态环保基础设施、能力建设和绿色产业发展项目;二是成立多国参与的"一带一路"绿色投融资担保机构,分担风险、撬动社会资本进入绿色领域;三是建立环境信息披露制度,基于"一带一路"生态环保大数据服务平台建设,开展环境信息披露。

促进环境产品与服务贸易便利化。提高环境产品与服务市场开放水平,鼓励扩大污染防治及处置技术和服务等环境产品和服务进出口,推动共建"一带一路"国家绿色产业发展。

5.1.4 加强能力建设,共同实施"一带一路"绿色能力建设活动

在公众参与方面,建议"一带一路"项目规划者采取以下措施。

大力促进共建"一带一路"国家民心相通。将绿色丝路使者计划打造成"一带一路"能力建设旗舰项目,通过开展环境管理人员和专业技术人员培训、政策内容指导等加强生态环境合作交流,分享中国生态文明和绿色发展的理念与实践。

支持和推动中国与共建国家环保社会组织交流合作。明确政府部门推动主体,引导和支持环保社会组织建立自身的合作网络;加大政府购买环保社会组织服务力度,设立支持环保社会组织走出去的专项合作资金;完善环保社会组织参与机制,制定环保社会组织参与的国际交流事项清单。

推动社会性别主流化,提升女性领导力。提升政策制定者和妇女群体的社会性别意识,推动社会性别意识纳入绿色"一带一路"政策制定与项目实施;推动中国国内生态保护相关机构在性别主流化方面能力建设,探索建立跨部门促进性别主流化的沟通机制;借助绿色丝路使者计划,组织共建"一带一路"国家生态环境领域女性官员、专家学者、青年学者等开展"提升女性绿色领导力"专题项目培训,并与"一带一路"合作伙伴分享实现性别主流化的方法与经验。

5.2 对接"一带一路"倡议与SDG15的政策建议

在上述绿色"一带一路"建设路线图的框架下,本节提出了"一带一路"倡

议与 SDG15 及《生物多样性公约》进行对接的政策方向。研究所提政策建议充分参考了国际上开展生物多样性保护时设立的主要目标及具体方法。主要目标包括建立和统一全球标准、制定可操作的风险管理策略、维护声誉及与利益相关方的合作关系。具体方法包括：在实践过程中与国际或国家承诺对标；使用不合格项目排除清单；开展包括生物多样性保护在内的环境影响评估；遵循缓解措施层级，避免对当地生态系统造成损害；纳入当地利益相关方。具体建议如下。

一是对接国际规则标准，鼓励采用较高环境标准。建议主动对接国际及国家承诺，推动绿色"一带一路"建设与《生物多样性公约》《联合国气候变化框架公约》等国际公约进行对接。此外，还应推动将"一带一路"与中国签署的其他生物多样性保护相关国际公约进行对接，包括《国际植物新品种保护公约》、《保护世界文化和自然遗产公约》、《濒危野生动植物种国际贸易公约》和《关于特别是作为水禽栖息地的国际重要湿地公约》，并发挥与《联合国气候变化框架公约》等气候相关公约的协同作用。

"一带一路"项目应符合所在国环境法律法规和标准，鼓励项目采用国际组织、多边金融机构实施的有关生态环境保护的原则、标准和惯例。根据中国银监会（CBRC）印发的《绿色信贷指引》，环境影响评估应确保"一带一路"项目符合项目所在国的生物多样性标准、中国和项目所在国均为缔约方的国际公约以及国家开发银行制定的可持续发展的投融资原则。此外，建议在"一带一路"框架下共同推动绿色价值链发展，实现国内国际双循环相互促进，共同推动绿色复苏。

二是聚焦环境影响，实施"一带一路"项目分级分类管理。依托"一带一路"绿色发展国际联盟正在开展的《"一带一路"项目绿色发展指南》研究，推动制定"一带一路"项目分级分类指南，重点关注项目在环境污染、生物多样性保护和气候变化等方面的影响，明确正面和负面清单，为共建"一带一路"国家及项目提供绿色解决方案，为金融机构提供绿色信贷指引。开展国家、城市和项目绿色发展试点示范，开展一批共建"一带一路"国家绿色发展案例研究和经验推广。《"一带一路"项目绿色发展指南》研究指出，分级分类管理应涵盖各项国际和国家承诺，满足所在国经济发展及环境保护需求，指导并协助项目在规划设计过程中将环境影响评估、生物多样性保护及影响减缓措施纳入其中。

三是完善政策工具，强化"一带一路"生态环境管理。政策工具包括以下四个方面。

（1）建议对"一带一路"重点行业、重点项目进行环境评估，建立常态化的环境管理机制，将环境污染、生物多样性保护、气候变化等环境因素作为评估的重要部分。充分考虑项目所在国的相关生态和社会经济条件，项目对动植物栖息地和生物多样性产生的直接、间接和累积影响，并将受到项目影响的利益相关方对生物多样性及动植物栖息地的重视程度考虑在内。

（2）充分运用绿色金融工具和环境风险分析方法，建立"生物多样性保护"治理与融资框架，发挥金融机构作用积极引导绿色投资。生物多样性丰富的热点地区主要位于发展中国家，难以设计和实施需要多年才能产生经济效益的长期项目，需要通过双多边合作开展生物多样性融资。过去十年，中国及全球范围内的生物多样性融资都出现了大幅增长，且在健全生态补偿机制、生态功能区转移支付、退耕还林补贴等方面积累了宝贵经验。生物多样性保护需要有利的政策环境，生态环境部应与国家发展改革委及其他行政机构共同努力，制定生物多样性保护的缓解策略，同时在与包括中国政府、项目所在国政府和其他相关各方与合作伙伴进行磋商后，为缓解、补偿和修复计划建立融资机制。此外，中国应在对共建"一带一路"国家的官方发展援助中增加环境援助的比例，加强生物多样性融资。

（3）将生态保护红线作为对接"一带一路"与SDG15的关键性工具，与共建"一带一路"国家交流共享在生态保护红线方面的良好实践经验，支持共建"一带一路"国家以生态保护红线为抓手，研究制定适合本国国情的类似生态保护红线的土地利用战略规划。

（4）注重建立"一带一路"利益相关方磋商机制，在生物多样性评估和管理方面保证相关方的有效参与和磋商。

四是加强协同机制，以基于自然的解决方案促进可持续发展目标有效衔接。充分发挥与SDG13气候行动等可持续发展目标的协同作用，建议逐步考虑减少在煤电等高碳行业的投入，防止高碳锁定，加强绿色环保低碳及可再生能源项目建设，更多建设一些环境可持续的绿色低碳项目。将绿色发展融入基础设施项目选择和实施管理，研究制定可持续基础设施建设运营指南。

附　录

附录1：第1章的支持性证据

表 A1-1　签订了"一带一路"备忘录的国家分布

区域	国　家
东亚	1. 蒙古国、2. 韩国
东盟10国	1. 新加坡、2. 马来西亚、3. 印度尼西亚、4. 缅甸、5. 泰国、6. 老挝、7. 柬埔寨、8. 越南、9. 文莱、10. 菲律宾
西亚17国	1. 伊朗、2. 伊拉克、3. 土耳其、4. 叙利亚、5. 约旦、6. 黎巴嫩、7. 以色列、8. 沙特阿拉伯、9. 也门、10. 阿曼、11. 阿拉伯联合酋长国、12. 卡塔尔、13. 科威特、14. 巴林、15. 希腊、16. 塞浦路斯、17. 埃及的西奈半岛
南亚8国	1. 印度、2. 巴基斯坦、3. 孟加拉国、4. 阿富汗、5. 斯里兰卡、6. 马尔代夫、7. 尼泊尔、8. 不丹
中亚5国	1. 哈萨克斯坦、2. 乌兹别克斯坦、3. 土库曼斯坦、4. 塔吉克斯坦、5. 吉尔吉斯斯坦
独联体7国	1. 俄罗斯、2. 乌克兰、3. 白俄罗斯、4. 格鲁吉亚、5. 阿塞拜疆、6. 亚美尼亚、7. 摩尔多瓦
中东欧16国	1. 波兰、2. 立陶宛、3. 爱沙尼亚、4. 拉脱维亚、5. 捷克、6. 斯洛伐克、7. 匈牙利、8. 斯洛文尼亚、9. 克罗地亚、10. 波黑、11. 黑山共和国、12. 塞尔维亚、13. 阿尔巴尼亚、14. 罗马尼亚、15. 保加利亚、16. 马其顿
西欧7国	1. 奥地利、2. 芬兰、3. 法国、4. 意大利、5. 卢森堡、6. 马耳他、7. 葡萄牙
北非5国	1. 阿尔及利亚、2. 利比亚、3. 毛里塔尼亚、4. 摩洛哥、5. 突尼斯
西非11国	1. 佛得角、2. 科特迪瓦、3. 冈比亚、4. 加纳、5. 几内亚、6. 利比里亚、7. 马里、8. 尼日利亚、9. 塞内加尔、10. 塞拉利昂、11. 多哥
中南非8国	1. 安哥拉、2. 喀麦隆、3. 乍得、4. 刚果共和国、5. 赤道几内亚、6. 加蓬、7. 纳米比亚、8. 南非
东非15国	1. 布隆迪、2. 吉布提、3. 埃塞俄比亚、4. 肯尼亚、5. 马达加斯加、6. 莫桑比克、7. 卢旺达、8. 塞舌尔、9. 索马里、10. 南苏丹、11. 苏丹、12. 坦桑尼亚、13. 乌干达、14. 赞比亚、15. 津巴布韦
拉美11国	1. 玻利维亚、2. 智利、3. 哥斯达黎加、4. 厄瓜多尔、5. 萨尔瓦多、6. 圭亚那、7. 巴拿马、8. 秘鲁、9. 苏里南、10. 乌拉圭、11. 委内瑞拉
加勒比地区8国	1. 安提瓜和巴布达、2. 巴巴多斯、3. 古巴、4. 多米尼克、5. 多米尼加、6. 格林纳达、7. 牙买加、8. 特立尼达和多巴哥
大洋洲9国	1. 斐济、2. 基里巴斯、3. 密克罗尼西亚、4. 新西兰、5. 巴布亚新几内亚、6. 萨摩亚、7. 所罗门群岛、8. 汤加、9. 瓦努阿图

注：东帝汶目前正在谈判加入东盟，下同。

表 A1-2 共建"一带一路"国家落实 SDG15 的进程

区域	国家	SDG15落实情况	SDG15变化趋势	区域	国家	SDG15落实情况	SDG15变化趋势
东亚	蒙古国		↗	西亚	伊拉克		→
东亚	韩国		→	西亚	科威特		.
东盟	新加坡		.	西亚	阿拉伯联合酋长国		
东盟	印度尼西亚		↓	西亚	沙特阿拉伯		→
东盟	马来西亚		→	西亚	叙利亚		→
东盟	柬埔寨		↓	西亚	以色列		↓
东盟	越南		↗	西亚	也门		↓
东盟	缅甸		↓	西亚	阿曼		
东盟	老挝		↓	西亚	土耳其		→
东盟	菲律宾		↓	西亚	巴林		.
东盟	泰国		→	西亚	黎巴嫩		
东盟	文莱			西亚	卡塔尔		
南亚	马尔代夫		.	西亚	伊朗		↓
南亚	印度		↓	西亚	埃及西奈半岛		→
南亚	阿富汗		↓	西亚	塞浦路斯		
南亚	不丹		→	西亚	希腊		↗
南亚	孟加拉国		↓	西亚	约旦		.
南亚	斯里兰卡		↗	中东欧	黑山共和国		↓
南亚	巴基斯坦		↓	中东欧	塞尔维亚		↗
南亚	尼泊尔		→	中东欧	波黑		→
中亚	土库曼斯坦		→	中东欧	马其顿		
中亚	哈萨克斯坦			中东欧	克罗地亚		
中亚	乌兹别克斯坦		→	中东欧	阿尔巴尼亚		↗
中亚	吉尔吉斯斯坦		→	中东欧	斯洛文尼亚		↑
中亚	塔吉克斯坦		→	中东欧	罗马尼亚		↑

附录

区域	国家	SDG15落实情况	SDG15变化趋势	区域	国家	SDG15落实情况	SDG15变化趋势
中东欧	斯洛伐克		↑	东非	吉布提		↓
	匈牙利		↑		马达加斯加		→
	立陶宛		↑		埃塞俄比亚		→
	爱沙尼亚		↑		坦桑尼亚		
	捷克		↑		肯尼亚		↓
	波兰		↑		苏丹		↗
	拉脱维亚		↑		卢旺达		→
	保加利亚		↑		莫桑比克		→
独联体国家	格鲁吉亚		→		乌干达		↗
	亚美尼亚		↓		赞比亚		
	乌克兰		→		布隆迪		↑
	摩尔多瓦				津巴布韦		↗
	俄罗斯				塞舌尔		.
	阿塞拜疆				索马里		→
	白俄罗斯		↑		南苏丹		↗
西欧	卢森堡		↗	中南非	南非		↗
	马耳他		.		安哥拉		→
	奥地利		↗		喀麦隆		→
	葡萄牙		↗		乍得		↑
	法国		↗		加蓬		↑
	芬兰		↑		纳米比亚		↑
	意大利		↑		刚果共和国		↑
北非	阿尔及利亚		→		赤道几内亚		.
	突尼斯		↗	西非	利比里亚		→
	毛里塔尼亚		.		佛得角		.
	摩洛哥		→		塞拉利昂		↑
	利比亚				马里		→

区域	国家	SDG15落实情况	SDG15变化趋势	区域	国家	SDG15落实情况	SDG15变化趋势
西非	塞内加尔		.	加勒比地区	牙买加		.
西非	冈比亚		.	加勒比地区	特立尼达和多巴哥		.
西非	加纳		.	加勒比地区	古巴		.
西非	科特迪瓦		↗	加勒比地区	多米尼克		↗
西非	尼日利亚		↗	加勒比地区	安提瓜和巴布达		.
西非	几内亚		↗	加勒比地区	巴巴多斯		.
西非	多哥		.	加勒比地区	多米尼加		.
拉丁美洲	乌拉圭		↓	加勒比地区	格林纳达		.
拉丁美洲	巴拿马		.	大洋洲	斐济		↓
拉丁美洲	智利		↓	大洋洲	瓦努阿图		.
拉丁美洲	圭亚那		.	大洋洲	新西兰		↓
拉丁美洲	厄瓜多尔		→	大洋洲	巴布亚新几内亚		.
拉丁美洲	萨尔瓦多		→	大洋洲	基里巴斯		.
拉丁美洲	哥斯达黎加		.	大洋洲	密克罗尼西亚		.
拉丁美洲	秘鲁		.	大洋洲	萨摩亚		.
拉丁美洲	苏里南		↗	大洋洲	所罗门群岛		.
拉丁美洲	玻利维亚		↗	大洋洲	汤加		.
拉丁美洲	委内瑞拉		↗				

图例：颜色   趋势箭头

绿色	实现目标	↑	进展顺利
黄色	存在挑战	↗	略有增加
橙色	较大挑战	→	停滞
红色	巨大挑战	↓	下降
灰色	无数据	.	无数据

附录2：第2章的支持性证据

表 A2-1　中国落实 SDG15 进展评估

可持续发展目标	开展的主要工作	评估指标	总体评估及变化趋势
15.1 到 2020 年，根据国际协议规定的义务，保护、恢复和可持续利用陆地和内陆的淡水生态系统及其服务，特别是森林、湿地、山麓和旱地	保护重要湿地及河口生态水位，保护修复湿地与河湖生态系统，建立湿地保护体系和退化湿地保护修复制度，推进湿地合理利用。推进陆地自然保护区法制体系建设，提高森林等自然资源的保护和利用水平。开展河湖健康评估，保护水生态系统	国家级水产种质资源保护区	●
		湿地公园数量	●
		地表水水质优良（Ⅰ～Ⅲ类）水体比例	●
15.2 到 2020 年，推动对所有类型森林进行可持续管理，停止毁林，恢复退化的森林，大幅增加全球植树造林和重新造林	开展大规模国土绿化行动，加强林业重点工程建设，完善天然林保护制度，全面停止天然林商业性采伐，保护和培育森林生态系统。完善退耕还林还草政策，探索开展政府购买、社会服务造林、护林工作机制	活立木总蓄积量	●
		天然林面积	●
15.3 到 2030 年，防治荒漠化，恢复退化的土地和土壤，包括受荒漠化、干旱和洪涝影响的土地，努力建立一个不再出现土地退化的世界	参与《联合国防治荒漠化公约》土地退化零增长目标设定的示范项目。推进荒漠化、石漠化、水土流失综合整治，预防土地沙化，不断拓展沙化土地治理范围，加强沙区生态保护和建设	重点生态工程区森林蓄积量	●
		重点生态工程区草原植被覆盖度	●
		沙化土地面积	●
15.4 到 2030 年，保护山地自然生态系统，包括其生物多样性，以便加强山地生态系统的能力，使其能够带来对可持续发展必不可少的益处	全面提升山地自然生态系统稳定性和生态服务功能，筑牢生态安全屏障。建设国家林木种质资源库，形成标准化的种质资源保存体系。科学优化森林公园建设管理体系，促进森林多样性资源的分享和利用	森林公园数量和面积	●
		活立木总蓄积量	●
		天然林面积	●
		国家生态保护资金投入	●
15.5 采取紧急重大行动来减少自然栖息地的退化，遏制生物多样性的丧失，到 2020 年，保护受威胁物种，防止其灭绝	实施生物多样性保护重大工程。强化自然保护区建设和管理，加大典型生态系统、物种、基因和景观多样性保护力度。加强生态系统保护与修复资金投入，开展全国大规模的物种资源本底调查工作。建立全国生物多样性观测网络体系	红色名录指数	●
		地球生命力指数	●

可持续发展目标	开展的主要工作	评估指标	总体评估及变化趋势
15.6 根据国际共识，公正和公平地分享利用遗传资源产生的惠益，促进适当获取这类资源	逐步建立健全遗传资源保护与惠益分享方面的法律法规，促进遗传资源的正当获取和公正、公平分享利用。提高生物遗传资源保护资金投入，参与遗传资源获取和利用的国际合作	遗传资源获取与惠益分享指标	…
15.7 采取紧急行动，终止偷猎和贩卖受保护的动植物物种，处理非法野生动植物产品的供求问题	认真执行《野生动物保护法》和加快完善《国家重点保护野生动物名录》，优化全国野生动物保护网络，强化野生动植物进出口能力，严厉打击象牙等野生动植物制品非法交易。修复和扩大濒危野生动植物栖息地，推进野生动物保护国际网络互联	/	/
15.8 到2020年，采取措施防止引入外来入侵物种并大幅减少其对土地和水域生态系统的影响，控制或消灭其中的重点物种	积极参与有关防控外来物种入侵的国际公约，完善外来入侵物种名单和相关风险评估制度	每10年新发现的外来入侵物种种数	●
		口岸截获有害生物的种数和批次	●
		发布的外来入侵物种风险评估标准的数量	●
15.9 到2020年，把生态系统和生物多样性价值观纳入国家和地方规划、发展进程、减贫战略和核算	要求各级地方政府结合本地区实际情况，因地制宜地做好生态环境和生物多样性保护工作，并将有关工作同本地区中长期发展规划有机结合	与生物多样性保护和可持续利用相关的部门政策数量	●
15.a 从各种渠道动员并大幅增加财政资金，以保护和可持续利用生物多样性和生态系统	加强协调，增加基础设施和能力建设所需资金	国家生态保护资金投入	●
15.b 从各种渠道动员资源，从各个层次为可持续森林管理提供资金支持，并为发展中国家推进可持续森林管理，包括保护森林和重新造林，提供充足的激励措施	推进多元化筹集资源战略，引导企业和社会公众更深入参与，形成森林管理的长效机制。在南南合作框架下帮助其他发展中国家开展技术培训，提升森林资源利用率和森林经营管理水平。指导中国企业在境外开展可持续森林经营与管理	森林生态效益补助资金	●

可持续发展目标	开展的主要工作	评估指标	总体评估及变化趋势
15.c 在全球加大支持力度，打击偷猎和贩卖受保护物种，包括增加地方社区实现可持续生计的机会	加强中国参加的国际贸易公约限制进出口物种的审查，严格《濒危野生动植物种国际贸易公约》证书管理。开展专项行动，遏制盗猎和非法贸易野生动物的犯罪势头。鼓励和引导野生植物人工栽培产业发展	查获的非法贩卖受保护物种的数量	…

●表示状况有改善；●表示状况在恶化；
…表示没有足够数据；/ 表示无对应指标，未开展评估。
资料来源：《中国履行生物多样性公约第六次国家报告》，2018 年。

表 A2-2　金融机构为客户制定的生物多样性保护方面的操作要求

	ADB	AFDB	AIIB	BNDES	CAF	EBRD	EIB	IDB	IFC	KFW	WB
依据项目对生物多样性的影响和风险程度对其进行评估和分类	X	X	X	X	X	X	X	X	X	X	X
评估基线条件	X	X	X		X	X	X	X	X		X
评估对生物资源的直接、间接、累积和诱发性影响及风险	X	X	X		X	X	X	X	X		X
考虑跨境影响	X	X	X			X	X	X			X
因生物多样性改变而导致的社会经济影响	X				X	X	X	X		X	X
采用战略环境评价	X	X	X				X	X	X		X
采用预防为主的措施或原则	X				X	X	X			X	X
考虑项目设计技术和各组成部分的替代方案	X	X	X		X	X	X	X			X
明确将环境缓解措施的成本纳入环境评估					X			X			
遵循减缓顺序		X	X	X	X	X	X	X	X		X
明确遵守国家法律和东道国的国际承诺	X	X	X	X	X	X	X	X	X		X

	ADB	AFDB	AIIB	BNDES	CAF	EBRD	EIB	IDB	IFC	KFW	WB
可选择使用国家和/或客户的保障体系代替金融机构的保障	X		X					X			X
聘请独立专家和咨询小组		X	X			X					X
在环境评估和项目实施过程中与利益相关方进行磋商	X	X	X		X	X	X	X	X	X	X
要求客户披露环境评估及管理计划	X		X			X	X		X		X
准备"生物多样性管理或行动计划"			X			X	X				
增强生物多样性	X	X	X		X	X	X		X		X
采用适应性管理以应对意外影响			X				X				
位于/影响关键栖息地的项目标准	X	X	X		X	X	X	X	X		X
位于/影响受法律保护和国际认可的地区的项目标准	X	X	X	X	X	X	X		X		X
位于/影响自然栖息地的项目标准	X	X	X		X	X	X		X		X
位于/影响恢复栖息地的项目标准	X					X		X	X		X
使用抵消补偿	X		X			X	X		X		X
生态系统服务管理		X			X	X	X		X		X
自然生物资源和可再生资源的可持续管理	X	X				X	X		X		X
控制外来入侵物种	X	X	X			X	X		X		X
转基因生物		X				X					
环境流		X*									
森林管理		X	X		X					X	
海洋管理		X	X		X					X	

	ADB	AFDB	AIIB	BNDES	CAF	EBRD	EIB	IDB	IFC	KFW	WB
保护传统知识和商业性活动		X			X				X	X	X
供应链管理		X				X	X		X		X
气候变化对生物多样性的影响		X				X	X				
完全不合格项目的清单	X	X		X	X	X	X	X	X	X	X

注：X 指影响水资源的项目。

资料来源：网页、官方政策文件及对所列国际机构中的工作人员的采访。

附录3：第3章的支持性证据

中国保护融资的详细描述：

加大生态功能区转移支付力度。自2008年中央财政设立国家重点生态功能区转移支付以来，国家不断加大对重点生态功能区的保护力度。2018年中央财政下达重点生态功能区转移支付721亿元，比上年增加94亿元，增幅达15%。与此同时，中国不断扩大国家重点生态功能区范围，在纳入国家重点生态功能区后，各地将获得相关财政、投资等政策支持，但必须严格执行产业准入负面清单制度。按照相关规定，纳入国家重点生态功能区的地区要强化生态保护和修复，合理调控工业化、城镇化开发内容和边界，保持并提高生态产品供给能力。

加强林业生态保护中央财政支持。2018年7月27日，财政部、国家林业和草原局发布《林业生态保护恢复资金管理办法》（以下简称《办法》），以加强和规范林业生态保护恢复资金使用管理，推进资金统筹使用，提高财政资金使用效益，促进林业生态保护恢复。《办法》规定，林业生态保护恢复资金是指中央财政预算安排的用于天然林资源保护工程（以下简称天保工程）社会保险、天保工程政策性社会性支出、全面停止天然林商业性采伐、完善退耕还林政策、新一轮退耕还林还草等方向的专项转移支付资金，2018年共计发放416.04亿元，其中黑龙江最多，高达85.95亿元。《办法》明确，林业生态保护恢复资金采取因素法分配。完善退耕还林政策现金补助标准为：长江流域及南方地区每亩退耕地每年补助125元，黄河流域及北方地区每亩退耕地每年补助90元。补助期限为：还生态林补助8年，还经济林补助5年。新一轮退耕还林还草补助标准为：退耕还林每亩退耕地现金补助1 200元，五年内分三次下达，第一年500元，第三年300元，第五年400元；退耕还草每亩退耕地现金补助850元，三年内分两次下达，第一年450元，第三年400元。

加大湿地生态保护修复中央财政支持力度。2013—2016年，中央财政共计安排50亿元支持中国湿地保护，此后继续通过林业改革发展资金支持湿地保护恢复。这些支持措施主要包括支持湿地保护与恢复、支持退耕还湿、支持湿地生态效益补偿。

持续推进草原生态保护补助奖励政策。自2011年，国家在内蒙古、新疆、西藏、青海、四川、甘肃、宁夏和云南8个主要草原牧区省区和新疆生产建设兵团，实施草原生态保护补助奖励机制并补贴136亿元，此后又将范围扩大到黑龙江等5个非主要牧区省的36个牧区半牧区县，覆盖了全国268个牧区半牧区县。近年来，国家在河北、山西等13个省（区）以及新疆生产建设兵团和黑龙江省农垦总局启动实施草原补奖政策，有力促进了牧区草原生态、牧业生产和牧民生活的改善，取得了显著成效。2018年，中央财政安排新一轮草原生态保护补助奖励187.6亿元，支持实施禁牧面积12.06亿亩、草畜平衡面积26.05亿亩，并对工作突出、成效显著的地区给予奖励，由地方政府统筹用于草原管护、推进牧区生产方式转型升级。其中，禁牧补助、草畜平衡奖励要求各地按照"对象明确、补助合理、发放准确、符合实际"的原则，根据补助奖励标准和封顶保底额度，做到及时足额发放。资金发放实行村级公示制，广泛接受群众监督。绩效评价奖励在可统筹支持落实禁牧补助和草畜平衡奖励基础工作的同时，要求各地用于草原生态保护建设和草牧业发展的比例不得低于70%，并因地制宜推进草牧业试验试点，加大对新型农业经营主体发展现代草牧业的支持力度。

积极推动自然资源统一确权登记试点。自然资源确权登记工作是推动自然资源资产产权制度改革的基础环节，而健全自然资源资产产权制度是中国生态文明制度建设的重要内容。2018年7月6日，自然资源部等7部委在北京召开了自然资源统一确权登记试点工作评估验收会，部分省份自然资源统一确权登记试点工作已顺利通过了评估验收。各试点地区以不动产登记为基础，以划清全民所有和集体所有之间的边界，划清全民所有、不同层级政府行使所有权的边界，划清不同集体所有者的边界，划清不同类型自然资源的边界"四个边界"为核心任务，以支撑山水林田湖草沙整体保护、系统修复、综合治理为目标，按要求完成了资源权属调查、登记单元划定、确权登记、数据库建设等主体工作，探索出一套行之有效的自然资源统一确权登记工作流程、技术方法和标准规范。

附录 4：第 4 章的支持性证据

4.1 中国生物多样性保护的政策执行

中国生物多样性保护的政策执行方式主要有立法、司法，科技创新和国际合作等。中国生物多样性保护法律体系初步建立，科技创新和国际合作深入推进。

国家立法包括作为国家根本大法的《宪法》，作为环境法律体系基本法的《环境保护法》和根据该法的基本原则所颁发的大量有关生物多样性保护的单行法律和行政法规（如《海洋环境保护法》《水法》《水污染防治法》《水土保持法》《渔业法》《森林法》《草原法》《野生动物保护法》《野生植物保护条例》《陆生野生动物保护实施条例》《植物新品种保护条例》《自然保护区条例》《风景名胜区管理条例》等）；另外如《水生动物保护条例》《水产资源繁殖保护条例》《专属经济区和大陆架法》和《渤海区渔业资源繁殖保护条例》等地方性有关生物多样性保护的法规；涉及湿地生物多样性的法律，包括《湿地公约》和《生物多样性公约》等多部行政法规。

地方立法包括《黑龙江省湿地保护条例》《甘肃省湿地保护条例》和《江西省鄱阳湖湿地保护条例》等 9 个省的地方保护条例。此外，还颁布了一系列行政法规，包括《自然保护区条例》《野生植物保护条例》《农业转基因生物安全管理条例》《濒危野生动植物进出口管理条例》和《野生药材资源保护管理条例》等。相关行业主管部门和部分省级政府也制定了相应的规章、地方性法规和规范。

中国加入的有关生物多样性保护的国际公约包括《生物多样性公约》《关于特别是作为水禽栖息地的国际重要湿地公约》《濒危野生动植物种国际贸易

公约》《保护世界文化和自然遗产公约》以及《人类环境宣言》和《里约环境与发展宣言》。涉及外来物种控制问题的相关法律主要有《进出境动植物检疫法》《动物防疫法》《海洋环境保护法》《家畜家禽防疫条例》等。对于新出现的转基因生物安全问题，国务院也于 2001 年紧急出台了《农业转基因生物安全管理条例》。这些法律的颁布和实施，对中国生物多样性的保护起到了重要的监督管理作用。

中国最高人民法院设立了环境资源审判庭，出台指导意见把生物多样性保护案件纳入专门化研究和审理范围，指导各级法院探索按照流域或者生态功能区跨行政区划集中管辖环境资源案件，统一环境资源案件的裁判标准，完善环境资源多元纠纷解决机制，为加强包括生物多样性在内的环境资源司法保护奠定坚实的基础。中国法院还充分发挥环境公益诉讼对生物多样性的保护功能，依法审理涉及湿地、林地、濒危植物、候鸟迁徙地等生物多样性保护的公益诉讼案件。在鄱阳湖之畔的千年古镇永修县吴城镇建立全国首个生物多样性司法保护基地，秉承严格执法、维护权益、注重预防、修复为主、公众参与等现代环境资源司法理念，旨在通过巡回审判、法制宣传等形式，在推进生态文明建设进程中，更加有效地发挥司法的服务和保障职能。

开展生物多样性基础调查、科研和监测，运用科技创新推动生物多样性可持续发展。有关部门先后组织了多项全国性或区域性的物种调查、科研与监测工作，建立了相关数据库，出版了《中国植物志》《中国动物志》《中国孢子植物志》以及《中国濒危动物红皮书》等物种编目志书。借鉴国际先进经验，开展试点示范，加强生物遗传资源价值评估与管理制度研究，探索建立生物遗传资源及传统知识获取与惠益共享制度，协调生物遗传资源及相关传统知识保护、开发和利用的利益关系。

提高公众参与意识，加强国际合作与交流。开展多种形式的生物多样性保护宣传教育活动，加强学校的生物多样性科普教育。建立和完善生物多样性保护公众监督、举报制度，完善公众参与机制。建立生物多样性保护伙伴关系，充分发挥民间公益性组织和慈善机构的作用，调动国内外利益相关方共同推进生物多样性保护和可持续利用。强化公约履行，积极参与相关国际规则的制定，引进国外先进技术和经验。

4.2　中国生物多样性保护的其他重要机构

- 中国生物多样性保护国家委员会。

2010 年，联合国大会把 2011—2020 年确定为"联合国生物多样性十年"，国务院成立了"2010 国际生物多样性年中国国家委员会"，召开会议审议通过了《国际生物多样性年中国行动方案》和《中国生物多样性保护战略与行动计划（2011—2030 年）》。2011 年 6 月，国务院决定把"2010 国际生物多样性年中国国家委员会"更名为"中国生物多样性保护国家委员会"，统筹协调全国生物多样性保护工作，指导"联合国生物多样性十年中国行动"。

- 中国科学院生物多样性委员会。

中国科学院于 1992 年成立了生物多样性委员会（BC-CAS），协调院内外生物多样性研究工作。其职责为：确定中国科学院生物多样性研究方针；制定中国科学院生物多样性研究长远规划及行动计划；审议观测、试验技术的规程和条例，审议各项组织管理制度，审定研究经费的分配方案；检查经费的使用情况及工作执行情况；审议学术交流与人才培养方案；制定国际、国内合作研究的规划和措施。中国科学院生物多样性委员会争取并负责实施世界银行贷款环境技术援助项目"生物多样性研究与信息管理"子项目。目前，已经建成 30 多个数据库，其中 25 个数据库、约 14 万条记录可以通过互联网查询。

4.3　金融机构保护的治理结构

- 欧洲复兴开发银行（EBRD）：借款人的任务是监督整体的管理、监测和报告。

- 国际金融公司（IFC）：国际金融公司与私营部门的借款人合作，建立一种"三角监管关系"：国际金融公司、客户及客户所在国的政府。在通常情况下，客户承担监测和报告的责任，但国家政府拥有对自然资源的管辖权或监督责任的情况除外。如果情况复杂或风险较高，则客户须借助外部专家的服务。

- 亚洲基础设施投资银行（AIIB）和拉丁美洲开发银行（CAF）：借款人的任务是进行监测和报告。金融机构还可以定期进行实地考察，并与项目实施者合力减轻已造成的破坏。

- 德国复兴信贷银行（KFW）：与借款人或客户达成监测和报告计划，然后由借款人或客户管理该计划。
- 亚洲开发银行（ADB）：借款人定期编写报告，而亚行则负责对报告进行尽职调查。亚行也会定期进行实地考察，并与实施者合力减轻已造成的破坏。
- 非洲开发银行（AFDB）：不定期地对那些具有生物多样性重大风险的项目进行独立审查，包括使用第三方审计员。一旦发现问题，该行将与借款人共同设计出结果可量化的行动计划，目的是增强地方监测和管理项目、减轻危害的能力。
- 美洲开发银行（IDB）和世界银行（WB）：金融机构监控合规情况，并监督报告。

表 A4-1 金融机构项目级申诉机制指导方针的共性

	AFDB	ADB	AIIB	EBRD	EIB	IFC	KFW	WB
机构定位								
独立，并由第三方进行监督	X							
可属于内部或外部（金融机构认为合适即可）				X				
资源								
应根据项目的风险和影响进行调整		X	X	X	X	X	X	X
应有足够的预算和人员配备					X			
设计与建设								
应与借款人/客户合作设计，以确保其合法性、可及性、可预测性和公平性	X							
应在项目开发过程中尽早建立				X				
流程								
应及时解决受影响人的问题	X	X		X	X	X		
整个流程应清晰、透明		X	X	X	X	X		
过程应可预测	X				X			
应注重两性平等		X	X					
文化上应契合		X	X	X		X	X	

	AFDB	ADB	AIIB	EBRD	EIB	IFC	KFW	WB
应不受操纵、胁迫或干扰				X				
应有公开的案例和结果记录	X		X					
应定期向公众报告其实施情况				X	X			
对申诉者的保障								
应保护申诉人不受恐吓/报复				X	X		X	
如果申诉者要求，应允许其匿名				X		X		
应对利益相关方免费	X				X	X		
受影响人群的各个阶层都应很容易获得		X	X					
客户应告知利益相关方机制的可用性			X	X		X		

注：AFDB：非洲开发银行；ADB：亚洲开发银行；AIIB：亚洲基础设施投资银行；EBRD：欧洲复兴开发银行；EIB：欧洲投资银行；IFC：国际金融公司；KFW：德国复兴信贷银行，德语名为Kreditanstalt für Wiederaufbau；WB：世界银行。

Part 2 英文部分

Green BRI and 2030 Agenda for Sustainable Development
—Aligning with Sustainable Development Goal 15 to Promote Global Biodiversity Conservation

绿色"一带一路"与2030年可持续发展议程
——对接2030年可持续发展目标 促进生物多样性保护

Executive Summary

The Belt and Road Initiative (BRI) promises to create new opportunities for shared growth among countries through policy coordination, connectivity, unimpeded trade, financial integration and people-to-people connections. It takes on new and deeper relevance amidst the global pandemic that has stricken the world. The fight against COVID-19 pandemic has made it abundantly clear that the global community is inescapably interconnected and needs stronger international collaboration through shared institutions and economic growth paths that are resilient, inclusive and sustainable. The BRI has the potential to make major contributions to these needs.

The BRI has significant potential to boost the incomes of BRI countries and the world at large. According to the World Bank, the BRI could increase trade in BRI countries by 9.7% and foreign direct investment (FDI) by 7.6%, which would lead to an increase in real income for Belt and Road economies by up to 3.4%. Increases of standards of living in the BRI countries also benefit the rest of the world, which according to the World Bank would grow by up to an additional 2.9% due to the BRI. These estimates stand in sharp contrast with similar estimates for the Trans-Pacific Partnership, which would have boosted the growth of its membership by just 1.1% and the rest of the world by 0.4%[1].

Alongside the significant benefits associated with major infrastructure financing, large infrastructure finance is also endemic to a set of sustainability-related risks, including biodiversity risk, and the BRI is no exception. Several early studies in China and abroad have shown that a number of the biodiversity risks common to infrastructure investment may also become common for the BRI. These studies show that the BRI may faced the challenge of losses in wildlife movement and mortality through habitat loss, the spread of invasive species, increases in illegal logging, poaching and fires; and cause deforestation through the construction of roads, power lines and power plants, and subsequent mining activity. For these reasons, it is important to incorporate eco-environmental risk mitigation and management into the "green BRI" framework to align it with the 2030 Agenda for Sustainable Development.

Chinese President Xi Jinping exhibited China's global commitment to biodiversity when he unveiled *the Beijing Call for Biodiversity and Climate Change* (Referred to as "in itiative") alongside French President Emmanuel Jean-Michel Frédéric Macron in late 2019. In the *call*, China and France pledge to lead by example to:

"Mobilize additional resources from all sources, both public and private, at the domestic and at the international level, towards both climate adaptation and mitigation; make finance flows consistent with pathways towards low greenhouse emissions and climate-resilient development, as well as for the conservation and sustainable use of biodiversity, the conservation of oceans, land degradation amongst others; ***ensure that international financing, particularly in the infrastructure field, is compatible with the Sustainable Development Goals (SDGs) and the Paris climate agreement***" (*China Daily*, 2019, emphasis added).

With the aim of fulfilling these commitments, this book examines how both China and the international community have learned over time to prevent and mitigate such risks. China's Ecological Red Line standards and analogous international practices offer a number of models that can be adapted to green the BRI with respect to biological diversity. The book includes further strategic principles for aligning the BRI with the Sustainable Development Goals (SDGs) and target of the *Paris Agreement* in general and establishing the green BRI Roadmap, i.e. China and the BRI participating countries are expected to mainstream green development through the "five connectivities" in building the Belt and Road, in order to jointly implement policies and measures for protecting eco-environmental and tackling climate change, as well as to support the international processes for environmental agreements such as *Convention on Biological Convention* (CBD) and the *United Nations Framework Convention* on Climate Change (UNFCCC). The green BRI roadmap links three frameworks: the green BRI, the 2030 Agenda for Sustainable Development and development goals of BRI participating countries. Specifically, this roadmap includes 4 major approaches. First, enhance policy communication. It is important to take the green BRI as an important practice of realizing SDGs and facilitating global environment governance reform with green development as the shared principle. Give full play to the role of BRI International Green Development Coalition (BRIGC) and other cooperation platforms. Second, enhance strategic alignment. It is suggested establishing the mechanism for linking Green BRI with the 2030 Agenda for Sustainable Development, actively promoting the alignment of environmental policies, planning, standards and technologies, and strengthening information sharing with the help of the BRI Environmental Big Data Platform. Third, improve project management. Establish and improve the mechanism for project management on green Belt and Road to further reinforce environmental management in BRI projects and prevent ecological and environmental risks from the development of BRI projects. Fourth, improve capacity building. It is recommended to jointly conduct green capacity building programs, such as the Green Silk Road Envoys Program, to create people-to-people bond in building green BRI.

Under the framework outlined by the above Roadmap for building a green BRI, with a special focus on Sustainable Development Goal 15 (SDG 15) and biodiversity conservation, more specific policy recommendations have been proposed to better align BRI, SDG 15 and *CBD*. This SPS recommends that China:

First, apply international norms and standards to facilitate the use of stricter environmental standards in BRI projects. It is recommended to actively align BRI efforts with the fulfillment of international and national commitments to international conventions, including the *CBD* and *UNFCCC*. **Second,** focus on environmental impacts and carry out assessment and classification-oriented management of BRI projects. It is recommended to boost the development of the guidance on assessment and classification of BRI projects, based on the on-going Joint Research on Green Development *Guidance for BRI Projects undertaken* by BRIGC, which could provide green solutions to BRI participating countries and projects. **Third,** improve policy instruments to prevent and control the eco-environmental risks related to BRI projects. It is recommended to carry out environmental impact assessment for key BRI sectors and projects and establish a regular environmental risk regulatory mechanism that incorporates environmental pollution, biodiversity conservation and climate change as important factors for assessment. It is important to make full use of green finance instruments and environmental risk assessment tools and take ecological redlining as a key instrument. **Fourth,** improve the coordination mechanism and facilitate effective linkage and alignment among different SDGs using Nation-based Solutions (NBS). It is necessary to create synergies with efforts for SDG 13 of Climate Action.

1 LINKAGES BETWEEN THE GREEN BELT AND ROAD AND THE 2030 AGENDA FOR SUSTAINABLE DEVELOPMENT

1.1 Background and Progress of Building the Green Belt and Road Initiative

1.1.1 The Background, Goal and Achievement of the Belt and Road Initiative

Since the financial crisis in 2008, the world has recognized the need to forge new sources and patterns of economic growth. In this context, the Belt and Road Initiative (BRI) was proposed as China's contribution to a comprehensive solution for sustainable development. Pursuing the principles of extensive consultation, joint contribution and shared benefits, the BRI promises to create new opportunities for shared growth and prosperity among countries through policy coordination, connectivity, unimpeded trade, financial integration and people-to-people connections. It takes on new and deeper relevance amidst the COVID-19 pandemic that has stricken the world, as it has become acutely clear that major international efforts like the BRI can help bolster cooperation against COVID-19 pandemics and other international challenges like financial crises, climate change and global biodiversity loss.

The accomplishments thus far have been impressive. From 2013 through 2019, cumulative commodity trade between China and countries along the Belt and Road, defined in the broadest terms, exceeded USD 7.8 trillion; direct investment to countries along the Belt and Road approximated USD 110 billion; and the value of new project contracts reached nearly USD 800 billion[2]. As estimated by the World Bank (2019)[3], implementing BRI projects will reduce the aggregate costs for trade among BRI participating economies by 3.5% and those for the trade between BRI participating economies with the rest of the world by 2.8%. By November 2019, the investment from Chinese enterprises in building economic and trade cooperation zones overseas in BRI countries amounted to USD 34 billion, creating tax revenue of over USD 3 billion and 320,000 local jobs[4]. According to the World Bank (2019), the implementation of the Belt and Road Initiative has the potential to raise real income gains raise incomes in BRI countries by 3.4% and increase global real income by up to 2.9% for the rest of the world. The BRI has been recognized by the United Nations as a solution for facilitating the implementation of the 2030 Agenda for Sustainable Development.

However, the BRI has even greater potential, specifically in the area of supporting biodiversity through high-quality infrastructure investment and global coordination. In April 2019, research findings and recommendation reports from the Advisory Council of the Belt and Road Forum (BRF) for International Cooperation (2019) highlighted that the Belt and Road Initiative and UN 2030 Agenda for Sustainable Development shared common ground in terms of facilitating cooperation, implementation instruments and measures among others, which could achieve greater synergy.

1.1.2 Progress of the Development of the Green BRI

Since its inception, building the Belt and Road into a pathway for green development has been the aspiration and expectation of the Chinese government as well as the shared goal of all participating countries. China has accelerated its progress in building an ecological civilization, making unprecedented efforts in recent decades. The concepts of "putting ecological progress in the first place" and "green development" have been widely accepted by Chinese society as a consensus, economic growth is shifting from a conventional model of "development first and green later" to high-quality development led by ecological civilization. By jointly building a green BRI with participant countries, China is creating a platform for countries to share and learn from one another the experience of green transitions and sustainable development. Over the past six years, China has been working closely with BRI participating countries in areas of environmental governance, biodiversity conservation and climate change mitigation and adaptation via bilateral and regional cooperation. It has witnessed positive and concrete results in building a green BRI and implementing the 2030 Agenda for Sustainable Development.

First, China has strengthened the BRI's top-level design and enhanced its cooperation mechanisms. In March 2015, the National Development and Reform Commission (NDRC), the Ministry of Foreign Affairs (MFA) and the Ministry of Commerce (MOFCOM) jointly issued their *"Vision and Actions on Jointly Building Silk Road Economic Belt and 21st-Century Maritime Silk Road."* The document proposes that China should promote ecological progress in conducting investment and trade, increase cooperation in ecological conservation, biodiversity protection and climate change mitigation and adaptation. In 2017, the Ministry of Ecology and Environment (MEE, then Ministry of Environmental Protection) issued the *"Belt and Road Ecological and Environmental Cooperation Plan"* and launched the *"Guidance on Promoting Green Belt and Road,"* which identified the roadmap for the development of a green BRI, together with the MFA, NDRC and MOFCOM.

As the BRI gradually unfolds, the green BRI framework is gaining a positive response from the international community. Currently, the MEE has signed nearly 50 bilateral and multilateral environmental cooperation agreements and has launched BRI International Green Development Coalition (BRIGC). The BRIGC was proposed by Chinese and international partners during the First Belt and Road Forum for International Cooperation (BRF), officially launched on the Thematic Forum of Green Silk Road of the Second BRF and listed as one of the sectoral multilateral cooperation initiatives and platforms in the Joint Communique of the Leaders' Roundtable of the Second BRF. The main goal of BRIGC is to promote international consensus, understanding, cooperation and concerted actions to achieve green development of the BRI. In December 2020, more than 150 Chinese and international organizations from over 40 countries have confirmed their partnership, including more than 70 overseas institutions such as government departments of BRI participating countries, international organizations, think tanks and businesses. Currently, BRIGC is actively promoting policy dialogues, thematic partnerships and champion projects. The flagship research on *BRI Green Development Report*, the Joint Research on the "*Green Development Guidance on BRI Projects*" (the "Green Light" System) and *the joint study on BRI Green Development Case Studies* have been launched.

Second, platforms and modes for cooperation have been enriched to be more pragmatic. China has expanded platforms for collaboration, including the China-Cambodia Environmental Cooperation Center and China-Laos Environmental Cooperation Office, which actively promote capacity building programs and champion projects. The Belt and Road Environmental Technology Exchange and Transfer Center (Shenzhen) was established to take advantage of the industrial resources of the area to promote innovative development and international transfer of environmental technologies. These platforms will facilitate environmental cooperation along the Belt and Road on regional and national levels. The BRI Environmental Big Data Platform (referred to as "the Big Data Platform") was officially launched. It has developed its own application (APP) for information updates, which helps to improve the "One-Map" system for integrated data services. With the help of information technologies, such as "Internet +" and big data, the Big Data Platform is designed to be an open platform for the exchange of ecological and environmental information through sharing and collaboration. It will provide environmental data support to BRI participating countries, including ecological environmental protection concepts, laws, regulations and standards, environmental policies and management measures, etc.

Third, China has promoted in-depth policy communication to build consensus on green development. China has made full use of existing international and regional cooperation mechanisms to share its vision, experience and achievements in ecological civilization and green development, through the UN Environment Assembly, CEEC Ministers' Conference on Environmental Cooperation and other international events. Meanwhile, the MEE is also engaged in opening up new channels for dialogue and communication. It held the Thematic Forum of Green Silk Road of the Second BRF for International Cooperation, organized sideline events on Green BRI during World Environment Day Celebrations, UN Climate Action Summit and China-ASEAN Environmental Cooperation Forum, and it sponsored more than 20 thematic forums each year on biodiversity conservation, climate change mitigation and adaptation, and eco-friendly cities with the attendance of more than 800 people from BRI participating countries and regions.

Fourth, these cooperation projects have borne fruit. For example, the Chinese government has established the Green Silk Road Envoys Program to promote capacity building in environmental governance in China and BRI participating countries. This program has trained more than 2,000 government officials, technological staff, youth and scholars from 120 BRI participating countries. According to the List of Deliverables of the Second BRF, the Chinese government will continue to implement the Green Silk Road Envoys Program, which expects to train 1,500 environmental officials from the BRI participating countries in the next three years. The Chinese government has also worked with relevant countries to jointly implement the Belt and Road South-South Cooperation Initiative on Climate Change to improve the capacity of BRI participating countries in addressing climate change and promote the implementation of the *Paris Agreement*. Moreover, China is also engaged in helping BRI participating countries in climate change mitigation and adaptation and energy transition, and promoting Chinese environmental technologies, standards, and low-carbon and energy-saving products in the international market through building low-carbon demonstration zones and organizing capacity building activities based on the reality and demands of BRI participating countries.

1.2 The Reason of Focus on SDG 15

In May 2019, the Intergovernmental Science-Policy Platform on Biodiversity and Ecosystem Services (IPBES) released *the Global Assessment Report on Biodiversity and Ecosystem Services*. The report evaluated the influence of biodiversity and ecosystem services on the economy, well-being, food security and life qualities. The report revealed that, over the past 50 years, the speed of biodiversity loss is unprecedented in human history. The top direct drivers for the most drastic biodiversity loss include changes in the use of land and sea, direct exploitation, climate change and invasive alien species; while values and behaviors such as demographic and sociocultural changes, economic and technological factors, as well as institutions and governance are considered as critical indirect drivers for biodiversity loss. Overall, 75% of the terrestrial environment has been severely changed by human behavior and activities. The pressures brought by the above drivers made it difficult to attain the related goals set by the *Convention on Biological Diversity* (CBD) and *the UN Framework Convention on Climate Change* (UNFCCC), unless more revolutionary actions are taken. Similarly, to realize relevant goals and targets in the 2030 Agenda, revolutionary changes from the status quo protection speed and measures have to be implemented.

The year 2021 will mark an important turning point. The 15th meeting of the Conference of the Parties (COP 15) to the CBD has been held in Kunming, China in 2021, with the theme of "Ecological Civilization: Building a Shared Future for All Life on Earth". COP 15 review the *Post-2020 Global Biodiversity Framework*, set up 2030 objectives and targets for the conservation of global biodiversity, formulate the strategy for the conservation of global biodiversity in a new decade (2021-2030), and launch the new course of post-2020 global biodiversity conservation.

The 2030 Agenda has highlighted the significance of biodiversity, with SDG 14 "Conserve and sustainably use the oceans, seas and marine resources for sustainable development" to deal with marine biodiversity and SDG 15 to address issues with terrestrial biodiversity. In this sense, CBD COP 15 could be considered as a key window of opportunity to speed up the attainment of biodiversity-related SDGs.

Built on the results of the first phase of the Special Policy Study (SPS) on Green Belt and Road and 2030 Agenda for Sustainable Development, this SPS, as the second phase of the series, will take a goal-by-goal and step-by-step approach to the alignment of BRI and biodiversity-related SDGs. This SPS will primarily focus on SDG 15 as the entry point and propose policy recommendations for COP 15 on how to encourage BRI participating countries to better implement SDGs with the help of BRI. Similar approaches and measures can be replicated in aligning BRI to SDG 14 and other SDGs pertinent to biodiversity in the future.

1.3 Progress of Countries along the Belt and Road in Implementing SDG 15

Progress is still lacking in achieving SDG 15 across BRI participating countries. *The Sustainable Development Report* 2019, published by the UN Sustainable Development Solutions Network (SDSN) and Bertelsmann Stiftung, evaluated progress among 193 countries in realizing SDG 13 (climate action), SDG 14 (life below water) and SDG 15 (life on land). It concludes that "trends on greenhouse gas emissions and even the trends on threatened species are moving in the wrong direction."

The SDSN assesses the progress of 139 BRI participating countries along the Belt and Road (listed in Table A1-1, Annex 1) towards realizing SDGs. The report selects five indicators to evaluate the implementation of SDG 15, including the mean area that is protected in terrestrial sites important to biodiversity (%), the mean area that is protected in freshwater sites important to biodiversity (%), the Red List Index of species survival, permanent deforestation (5 years average annual %) and imported biodiversity threats (per million population).

SDSN finds particularly strong challenges in the geographic regions most closely associated with Belt and Road corridors: the Association of Southeast Asian Nations (ASEAN) as well as West and South Asian countries. These results are discussed below. The detailed evaluation results are shown in Figure 1-1 below and Table A1-2 in Annex 1.

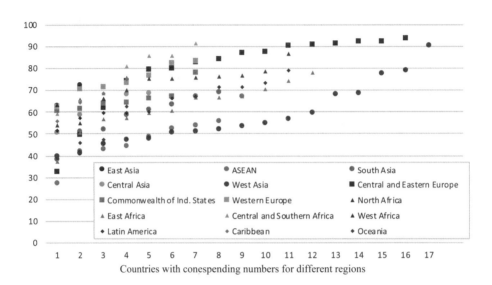

Figure 1-1　Score of BRI participating Countries on SDG 15

Note: 1. Circles indicate Asia, squares indicate Europe, triangles indicate Africa, and diamonds indicate other regions.
2. The horizontal coordinate is the corresponding numbercel country for the different regions in Table A1-1 of Amnes 1.

From the perspective of implementing SDG 15, SDSN finds that only four Central and Eastern European countries out of 139 countries have realized "Goal Achievement" of SDG 15: Poland, Hungary, Romania and Bulgaria. The implementation of SDG 15 in Central and Eastern Europe is generally better than in other regions. For countries in other regions, there are various degrees of risks in the implementation of SDG 15. "Major Challenges" exist for three ASEAN Member States (Malaysia, Indonesia and Viet Nam), four in South and West Asia (Afghanistan, Iraq, Turkey and Syria), four in East Africa (Djibouti, Madagascar, Seychelles and Somalia) and four in Oceania (Fiji, Micronesia, the Solomon Islands and Vanuatu).

Regarding the time sequence of implementing SDG 15, SDSN finds once again that Central Eastern European countries exhibit better performance than other regions. Ten out of 16 Central and Eastern European countries are on track or maintaining achievement, four countries show a moderately improving trend and two countries show stagnation in their work. ASEAN Member States and countries in South Asia are the main areas facing challenges. The scores of SDG 15 in half of the 10 ASEAN Member States are decreasing, while two countries are in stagnation. Four countries out of eight in South Asia demonstrat a declining trend in implementation performance. Most countries in Central Asia and Commonwealth of Independence States (CIS) reveal stagnation in implementation, including five Central Asian countries and five out of seven CIS countries.

From the perspective of specific indicators, the most impactful indicator for ASEAN and South Asian countries in implementing SDG 15 is the Red List Index. Following a time sequence, the performance on this indicator in ASEAN and South Asian countries exhibites a decreasing trend. In addition, for ASEAN Member States, permanent deforestation also brings tremendous risks in implementing SDG 15. Specific results are listed in Annex 1.

1.4 Benefits and Biodiversity-Related Risks of BRI

The BRI has the potential to close major infrastructure gaps, accelerate regional integration and increase economic growth in a manner that advances progress towards the SDGs. Indeed, there is certain evidence that after just a few years the BRI is contributing to the achievement of some of these goals. Any large-scale development effort also has potential risks, the key to the BRI's success will be to maximize the potential benefits while minimizing the potential risks. One such risk is the biodiversity decline that is often associated with major infrastructure investments in ecologically fragile areas with insufficient risk assessment in advance or risk management in operation. When accentuated, biodiversity loss could even jeopardize the economic returns of infrastructure investments.

Chinese President Xi Jinping exhibited China's global commitment to biodiversity when he unveiled *the Beijing Call for Biodiversity and Climate Change* (Referred to as "initiative") alongside French President Emmanuel Jean-Michel Frédéric Macron in late 2019. In the *call*, China and France pledge to lead by example to:

"Mobilize additional resources from all sources, both public and private, at the domestic and at the international level, towards both climate adaptation and mitigation; make finance flows consistent with pathways towards low greenhouse emissions and climate-resilient development, as well as for the conservation and sustainable use of biodiversity, the conservation of oceans, land degradation amongst others; ***ensure that international financing, particularly in the infrastructure field, is compatible with the Sustainable Development Goals (SDGs) and the Paris climate agreement***."

This SPS is intended to conduct evidence-based research in order to formulate a framework of policies that will help the BRI be compatible with SDG 15. This section outlines the potential and realized benefits of the BRI and the potential biodiversity loss risks associated with the BRI.

1.4.1 Benefits of the BRI

The world community faces a financing gap of 2.1% of global GDP annually to 2030 in order to provide the infrastructure that is needed to meet the SDGs[5]. The China-led BRI has the potential to take a leading role in closing those gaps in a manner that is aligned with the SDGs. According to estimates from the World Bank (2019), the transport corridors of the BRI will significantly increase economic growth in BRI countries. New transport corridors can increase the speed and efficiency of trade routes, connect isolated human settlements and create better access to markets by facilitating the transportation of goods, services and people across the world. When infrastructure is completed, there are boundless possibilities for "spillover effects" where new forms of economic activity arise, which would not achiere without the infrastructure investment[6].

The BRI has significant potential to boost the incomes of BRI countries and the world at large. According to the World Bank, the BRI could increase trade in BRI countries by 9.7% and foreign direct investment (FDI) by 7.6%, which would lead to an increase in real income for Belt and Road economies by up to 3.4% and by up to an additional 2.9% for other countries. In contrast, estimates for the Trans-Pacific Partnership (TPP) show that TPP would have boosted the growth of its membership by just 1.1% and the rest of the world by 0.4%[1]. The BRI then may have the largest potential to boost economic prosperity among participant countries and across the world.

These benefits are already being realized. Dreher et al. (2017)[7] looked at the impact of China's overseas projects financed by the China Development Bank, Export-Import Bank of China and other Chinese financial institutions on economic growth in 138 countries. The authors found that Chinese-financed projects could have a 0.7% increase on average in economic growth two years after the project is committed.

1.4.2 Biodiversity Risks and the BRI

Alongside the significant benefits associated with major infrastructure financing, large infrastructure finance is endemic to a set of sustainability-related risks, including biodiversity risk, and the BRI is no exception. In a recent article in the journal *Nature Sustainability*, biodiversity experts noted that a number of the biodiversity risks that are common with respect to infrastructure investment may also become common for the BRI. The authors express concern that "the expansion of transportation networks will increase habitat loss, the overexploitation of resources and the degradation of surrounding landscapes."[8] In particular, the authors noted that the BRI could face the potential problems of losses in wildlife movement and mortality through habitat loss, spread invasive species, increase in illegal logging, poaching and fires, and cause deforestation through the construction of roads, power lines and power plants and subsequent mining activity. The authors also note that "such impacts, which are already high in some regions, will degrade ecosystem services, possibly pushing some ecosystems beyond tipping points, where small negative changes can lead to abrupt changes in ecosystem quality and functionality"[8].

A handful of studies have already identified some of the potential biodiversity risks of the BRI. In a recent article in *Conservation Biology*, Hughes (2019) spatially located proposed road and rail projects of the BRI (defined as those along the BRI corridors) and examined the extent to which such projects will be proximate to Key Biological Diversity Areas (KBAs) across the world.

The earliest study was conducted by the World Wildlife Fund for Nature (WWF). According to WWF's analysis, BRI corridors in Eurasia overlap with the range of 265 threatened species including 39 critically endangered species and 81 endangered species, with 1,739 Important Bird Areas or KBAs and 46 biodiversity hotspots or Global 200 Ecoregions. WWF finds the potentially most impacted areas to be the China-Indochina Peninsula Economic Corridor, the Bangladesh-China-India-Myanmar Corridor and the China-Mongolia-Russia Economic Corridor. A background study for the World Bank analysis discussed above came to similar conclusions. The China-Indochina Peninsula Economic Corridor and China-Mongolia-Russia Economic Corridor are facing the highest risks of biodiversity loss due to deforestation[10].

To appropriately address these risks, China's development finance institutions, which provide the bulk of the lending necessary for BRI projects to move forward, can institute safeguards that work with BRI participating countries to screen, assess, and oversee the operation to ensure best practices. A 2020 study in *Nature Sustainability* evaluated policies in financiers associated with the BRI: 35 Chinese and 30 international institutions. The authors found that only 17 of these lenders require biodiversity impact mitigation and only one of those is Chinese: the China-ASEAN Investment Cooperation Fund[11]. As a result, China faces potentially severe challenges in establishing cooperative mechanisms to oversee and mitigate biodiversity risks associated with specific BRI projects. This SPS explores lender safeguards and biodiversity risks mitigation in more detail below, in order to explore the potential for advancement in these areas.

Biodiversity loss also reduces economic well-being. A study published in the journal *Global Environmental Change* found that between 1997 and 2011, the world economy lost between USD 4 trillion and USD 20 trillion per year in ecosystem services from land cover changes[12]. A 2019 World Bank study examining the economic impact of ecological conservation efforts in Kenya shows that biodiversity management can make the difference between infrastructure projects having positive or negative economic impacts because of impacts on ecosystem services for surrounding communities[13].

Risks to biodiversity clearly carry potential impacts for human communities, but those impacts can manifest differently across gender lines, which can severely curtail the effectiveness of conservation planning if it is not taken into account. In many rural, poor settings, biodiversity loss impacts women to a greater extent than men, especially in communities where women are tasked with collecting water, firewood and wild foods, which is common in developing countries globally[14-15]. If forests and riverine ecosystems are damaged, their tasks become more onerous, requiring farther travel in often insecure areas.

In addition to decreasing women's living standards, these gender-based impacts of biodiversity loss can have a compounding effect on the biodiversity loss itself by curtailing women's ability to fulfill their traditional role as local biodiversity stewards. In many rural areas around the world, women protect agro-biodiversity for their communities through the maintenance of household or communal gardens, while men are tasked with paid labor in agribusiness monocrop production[16]. In these communities, survival depends on both types of labor. During droughts, floods, or other natural disasters, the resilience of heirloom food crop varieties becomes especially important. These gardens are also more dependent on soil and water health, as chemical inputs are costly.

Thus, biodiversity conservation supports gender parity, which in turn further supports conservation. Development projects can support a virtuous cycle, or alternately, can initiate a descent into a vicious cycle in which forest and river biodiversity loss is compounded by their impacts on women, the traditional caretakers of crop biodiversity.

These same impacts can be seen in poorly designed conservation projects, in which women are unable to access the forests and waterways they traditionally visit for their sustainable gathering work[17]. Even though project planners may hope that their efforts preserve biodiversity, by not taking into account the gender-based impacts of their programs, they may limit the biodiversity benefits, as women must shift their time from managing crop biodiversity to traveling greater distances for gathering basic household needs.

Unfortunately, these gender-based risks to biodiversity management can be difficult for planners to detect if they are not specifically looking for them. As Lu et al. (2018)[18] point out, in contexts where women do not customarily participate in public discussions, the impact they face from development proposals may go undetected. In these settings, even projects that rely on community participation will miss input from women if they are not specifically prioritized, leaving planners and funders vulnerable to the risk of biodiversity loss[19-21].

If the BRI does not develop and institutionalize a strategic set of appropriate policies and standards to mitigate the biodiversity risks, it could encounter financial, social, environmental and political risks as well that may further erode the maximum potential of the BRI. Fragile ecosystems can jeopardize the integrity of infrastructure projects, reduce financial rates of return and accentuate debt-driven macroeconomic stress in host governments and on the balance sheets of Chinese financiers. Furthermore, increased degradation of biodiversity can lead to social conflict and reputational risks that can also threaten the geo-political relationships that are so important to the BRI as well. For the above reasons, it is important to control biodiversity risks associated with the BRI.

1.5 The Need for Biodiversity Policy in the BRI

Concrete policies for biodiversity conservation will be key to maximizing the potential benefits of the BRI, with the aim of maximizing the benefits of the BRI, the rest of this SPS report surveys best practices across China and the globe with respect to project finance and biodiversity in order to draw lessons for a coherent set of policies that China could adopt for the BRI moving forward. Section 2 of the SPS report surveys and assesses policies and standards for biodiversity conservation in China and by international institutions. Section 3 examines different biodiversity finance (Biofin) policies in China and abroad. Section 4 surveys potential Chinese and international governance structures that may be appropriate for incorporating biodiversity into the BRI. Finally, Section 5 distills a set of policy recommendations that Chinese institutions could adopt to align the BRI with SDG 15.

2 AN ANALYSIS OF RELEVANT POLICIES AND STANDARDS ON SDG 15

2.1 Research and Evaluation of China's Experience

2.1.1 Biodiversity Conservation in China

China is among the world's megadiversity countries, yet its biodiversity is seriously threatened. To strengthen biodiversity conservation, China has been conducting biodiversity surveys, assessments of endangered categories of ecosystems and species and in-situ and ex-situ conservation, as well as developing policies and regulations on biodiversity conservation. Assessment of China's progress in implementing SDG 15 is displayed in Table A2-1 in Annex 2.

In terms of in-situ and ex-situ conservation, China has established a natural protected area system pivoting on national parks and also included nature reserves, scenic areas, forest parks, geographic parks, wetland parks and cultural and natural heritage sites among others. To supplement the natural protected areas, China has also established key ecological function zones and priority areas for biodiversity conservation. Currently, China has more than 10,000 protected areas, including national parks, nature reserves, forest parks, scenic areas, geographic parks, wetland parks, drinking water sources and so on, covering 18% of the national land territory. At the same time, China has proposed an ecological function zoning scheme that consists of large-scale ecological function zones of different levels (including national key ecological function zones, important ecological function zones, bio-sensitive zones and vulnerable zones), which has played a significant role in protecting biodiversity and safeguarding national ecological security. However, even with these measures in place, China has still witnessed severe ecosystem degradation and accelerated biodiversity loss due to a lack of clear identification of natural protected areas' boundaries. The drawing of ecological redlines could identify areas with unique ecological functions, which must be strictly protected in order to realize centralized management of the eco-space.

2.1.2 Practices of Ecological Redlining in China

1. The Drawing and Management of Ecological Redlines

In October 2011, the State Council of China released the *Opinions on Strengthening Major Environmental Protection Work* to put forward ecological redlining for the first time. The document articulated the drawing of ecological redlines in major ecological function areas, sensitive areas and vulnerable areas for permanent conservation. In February 2017, the General Office of the CPC Central Committee and the General Office of the State Council jointly issued and circulated *Opinions on Drawing and Strictly Following Ecological Redlines*, which established the framework, basic principles and overall goal of delineating and observing the ecological redlines. The release of this document represented a new phase of accelerated development of the ecological redline system in China.

2. The Development of Scientific Methodology for the Drawing of Ecological Redlines

Scientific assessment is necessary before drawing ecological redlines. The aim of this step is to identify the spatial distribution of areas with critical ecological functions (such as water conservation, biodiversity protection, and water and soil preservation) and areas sensitive or vulnerable to water loss and soil erosion, desertification and salinization. The next step is to conduct a spatial mapping analysis of the two categories of areas and draw a redline for ecological protection that encompasses all development-prohibited areas at national and provincial levels and other protected areas in need of strict protection.

The design of ecological redlines aimed to bring almost all rare and endangered species in China and their habitats under protection, with due consideration to China's own reality. Ecological redlining doesn't equal identifying new protected areas, but rather, constructing and optimizing the systems for ecological protection with a more scientific, comprehensive and systematic approach. It could turn existing protected areas into an integrated ecological protection system that is easy to manage. It contains both established protected areas of all kinds and areas that lack protection.

3. The Establishment of the System for Delineating and Observing Ecological Redlines

In drawing the ecological redlines, the national government develops technical guidelines for provincial governments to decide the areas to be covered autonomously. Based on the *Methods for the Management of Ecological Redlines* issued by the Central Government, provincial governments develop their own methods with reference to local reality with detailed regulations on environmental access, the sustainable utilization of resources, ecological conservation and restoration, compensation for ecological protection and assessment and evaluation. Governments of all levels should take the responsibility of managing and regulating the ecological redlines.

4. Significant Effects have been Achieved

In January 2018, the State Council approved the redline drawing plans from 15 provinces (autonomous regions and municipalities), including Beijing, Tianjin, Hebei provinces and municipalities in the Yangtze River Economic Belt, and Ningxia. All these plans have been promulgated and implemented. In October 2018, the MEE and the Ministry of Natural Resources of China organized review meetings, principally approving the plans of drawing ecological redlines in 16 other provinces (autonomous regions and municipalities). The areas and sites covered by ecological redlines should be specified and demarcated after surveys. Still, based on the drawing plan, the ecological redline areas nationwide account for one-third of the national territory. Major ecological land within the redline boundaries, including forests, grasslands and wetlands, accounts for 55% of the major ecological land nationwide. The natural protected area system pivoting on national parks has covered more than 18% of China's national land territory, surpassing the ratio of 17% set out by the 2020 Aichi Biodiversity Targets. The wild population of certain rare and endangered species such as the giant panda, crested ibis and Tibetan antelope, has steadily increased. The major ecological land protected by the redlines covers the catchment areas of the Yangtze River, the Yellow River and the Pearl River, among other major rivers at and above Category III in China, as well as all biodiversity-rich areas identified at the national level and the vast majority of biodiversity-rich areas defined at provincial levels. Redlining has also protected most river and lake water sources as well as some underground water sources, all the distribution areas of species on the List of Wildlife under Special State Protection, as well as the areas where protected fauna and flora are mostly distributed.

2.1.3 The Experience of China in Biodiversity Conservation through the Ecological Redline Policy (ERP)

Ecological redlines help with biodiversity conservation through bringing areas with rich biodiversity and of importance under protection. In this way, habitats within the ecological redline can be preserved and restored, while in-situ and ex-situ biodiversity conservation can be realized.

1. The drawing of ecological redlines should be scientific and rational

An integrated and systematized approach of natural ecology should be taken to draw ecological redlines. Scientific assessment is needed for the identification of different areas based on the importance of ecological functions and the sensitivity and vulnerability of eco-environment. Areas within the ecological redline include all development prohibited areas on the national and provincial levels and other protected areas where strict protection is necessary.

2. Human activities should be strictly controlled in areas protected by ecological redlines

In terms of functional positioning, ecological redlines are of great significance to maintaining ecological equilibrium and supporting sustainable economic and social development; Areas within ecological redlines are land with critical ecological functions, the use of which must be strictly controlled; In terms of conservation, ecological redlines represent the critical point and baseline for safeguarding ecological security; Areas within ecological redlines should never be allowed to see degradation in their function, shrinking in their size or change in their nature. In principle, ecological redlines should be managed the same way as "development prohibited" areas, with all development activities not in line with the function positioning of the areas being strictly prohibited.

(1) The management of protected areas, including national parks, nature reserves, scenic areas, forest parks, geographic parks, world natural heritage sites, wetland parks and drinking water sources, should follow related laws and regulations.

(2) For other areas within ecological redlines, the following human activities are prohibited: mining activities; land reclamation, sand quarrying and other activities that may destroy coastlines; large scale agricultural activities, including wasteland reclamation, animal husbandry of scale and fishing; textile, printing and dyeing, leather manufacturing, paper-making and other manufacturing activities; real estate development; the construction of passenger and freight stations, ports and airports; coal-fired power and nuclear power generation and hazardous articles warehousing; the production of products with heavy pollution and high environmental risks listed in the *Comprehensive Directory of Environmental Protection* (2017), and production and operation activities with high environmental risks identified by *Measures for the Administration of Compulsory Liability Insurance for Environmental Pollution*.

3. Ecological restoration and ecological compensation should be conducted in areas protected by ecological redlines

(1) Conducting ecological restoration

China will soon develop plans for ecological conservation and restoration within ecological redlines that prioritize the protection of sound ecosystems and major habitats, restore damaged ecosystems, establish ecological corridors and sites and improve the integrity and connectivity of ecosystems. Ecological restoration in areas protected by ecological redlines is identified as an important component in the protection and restoration of ecosystems, including mountains, waters, forests, lakes, grassland and sand. The government is determined to effectively provide financial resources for ecological conservation and restoration by coordinating funding channels for various conversation and restoration projects within ecological redlines, such as programs on water and soil conservation, natural forest conservation and comprehensive improvement of land and resources. Ecological restoration within marine ecological redlines will be conducted, based on the principle of integrated governance of the land and the sea, with special emphasis on the comprehensive management of estuaries, littoral zones, islands and polluted waters.

(2) Introducing an ecological compensation mechanism integrating government funding and funding from other sources

Governments of all levels should increase their funding for areas protected by ecological redlines. Local governments are encouraged to launch fiscal, credit, financial and tax policies to facilitate the implementation of ecological redline policies and establish ecological compensation mechanisms.

Local governments should develop diversified investment and financing mechanisms guided by the government with extensive public engagement to pool in resources from all sides. Governments are also encouraged to launch pilot programs on payment for ecosystem services and develop market-based mechanisms to realize the value of ecological products.

4. Integrated monitoring should be developed and continuously improved for ecological redlines

It is important to access real-time statistics, improve the capability of the integrated analysis and application of monitoring data, be informed of the composition, distribution and dynamic change of ecosystems protected by ecological redlines, and keep track of human interference. Administrative decisions should be made in a scientific way with illegal acts being checked and handled in time.

5. An accountability system should be established for safeguarding ecological redlines

(1) Strengthening supervision for law enforcement

With the establishment of enforcement mechanisms for ecological redlining, regular supervision and inspection of law enforcement should be conducted to identify and punish illegal acts damaging ecological redlines. Cases must be inrestigated, violation must be accountable.

(2) Establishing an assessment mechanism

Assessment of the performance of local governments in implementing ecological redlines should be conducted. The results of the assessment shall be a reference in determining the political achievements of local governments.

(3) Strengthening the accountability system

Government officials whose decisions/actions cause severe damage to the ecological environment and resources shall be held accountable in all cases, regardless of their current positions.

(4) Launching an incentive mechanism

Rewards should be given to organizations and individuals that have outstanding performance in protecting ecological redlines. It is recommended that personnel be assigned for the promotion of ecological redlines to improve the engagement of local residents.

(5) Improving information transparency and public engagement

Governments should release information concerning ecological redlines, including their distribution and adjustment, to safeguard people's right to know, participate and supervise and give full play to the role of the media, non-governmental organizations (NGOs) and volunteers in promoting ecological redlines.

2.2 International Standards Related to SDG 15

Adopting a set of harmonized standards for SDG 15 across the BRI can help minimize the risk and maximize the benefits and legitimacy of all actors involved, through bolstering environmental and social risk management (ESRM). This section reviews the international standards related to SDG 15 as practiced by the major multilateral financiers of infrastructure, integration and development finance across the world, in two sections. First is a short note about the benefits of putting standards in place. Second is a comparative analysis of some of the major policies practiced by international actors.

Over the last few decades, environmental assessment and oversight systems have proliferated in the realm of international finance and investment. This section identifies the international actors that serve as peers for the Chinese financial institutions most active in BRI project finance and provides a survey of common practices among them. BRI projects predominantly receive financing through Chinese official entities such as the Silk Road Fund, the China Development Bank and the Export-Import Bank of China, though not exclusively so[22]. Thus, the international equivalent for the sake of environmental governance of cross-border infrastructure development is the cohort of multilateral development finance institutions (DFIs) that have been traditional sources of support for BRI participating countries.

2.2.1 Benefits of Developing Green Standards and Safeguards Across the BRI

Developing green standards can ensure that the BRI is calibrated to the SDGs while bringing benefits to virtually all of the stakeholders in the BRI. High-level or best-in-class environmental standards should thus take into account the preferences of Chinese and the other multitude of stakeholders engaged in the BRI to ensure that the BRI can provide public goods to the global economy as a whole.

Table 2-1 Benefits of Standardizing the SDGs in the BRI

Chinese actors	Expansion of markets
	Greater project effectiveness
	Prevention from default risk
	Prevention and mitigation of environmental and social risk
	Prevention and mitigation of reputational risk

Host countries	Improved management of fiscal resources
	Better management of natural resources
	Strengthening of institutional capacities
	Prevention and mitigation of environmental and social risk
	Prevention and management of reputational risk
Local communities	Reduced likelihood of social conflict
	Enhanced voice and ownership
	Reduced vulnerability
	Improved livelihoods
Global	Equitable use of resources
	Enhancement of global public goods
	Interconnectivity and global growth
	Leadership and legitimacy

Source: Authors' adaptation base on World Bank (2010) and China Development Bank-UNDP (2019).

Standards can also increase project performance and profitability of projects. For example, in 2018 the International Finance Corporation (IFC) found that establishing standards across each of the common norms noted above were correlated with strong financial performance (measured by return on assets and return on equity) and financial risk ratings in 656 IFC projects representing USD 37 billion. Risk instruments based on debt sustainability analysis (DSA) can help ensure that Chinese actors do not have to bear the risk of default on projects. While full assessments of the costs and benefits of ESRM are hard to quantify, the Independent Evaluation Group (IEG) of the World Bank (an independent monitoring group) conducted an assessment of the costs and benefits of ESRM in 2010 and concluded that benefits from the "environmental safeguards far outweigh the incremental costs"[23]. Weighing risks and benefits from a sample of bank projects, the World Bank found that most sensitive projects yielded "low cost - low benefits or high cost - high benefits for recipient countries." In the same IEG survey mentioned above, the World Bank also found that over half of the "task team leaders surveyed reported that the Bank's safeguards increased acceptability of the project among beneficiaries, and the safeguard policies also increased acceptability among nearly 30% of co-financiers" [23].

> **Box 2-1 Case study: Incorporating ESRM into Chinese Mining Enterprises in Peru**
>
> Chinese financiers, firms and the government can benefit substantially from establishing a set of harmonized standards around these common norms. First, these tools can help Chinese banks and firms expand and maintain market share overseas. China's experience in Peru is a case in point. Because of a lack of ESRM on the part of Chinese investors and the Peruvian government, China's first foray into Peru was a costly one. Chinese firms struggled to work with workers and local communities over worker health and safety, emergency preparedness and biodiversity concerns, though some of the issues were actually due to a lack of enforcement of host country systems rather than the Chinese firm, Chinese firms in general suffered reputational damage. Indeed, it became more difficult for Chinese firms to win contracts for mining and exploration in that country because of the perception that Chinese firms and financiers did not have proper risk management strategies. Later, Chinese copper firms devised significant ESRMs and participated in stakeholder consultations during the design stage. Such activity helped get market access and enhance China's reputation rather than worsen it. Indeed, when an accident did occur, ESRM plans allowed the company and host country to respond in such a way as to mitigate the worst damage[24-25].

Standards can also benefit local communities close to projects. Engaging with local laborers and communities about a project beforehand can help identify concerns. In Bolivia, Chinese tin companies took part in a prior informed consent engagement with local communities which didn't fully support the location of the tin company. Bolivia found another community more suited and equipped for the project, likely deferring social conflicts that would have hurt the companies' business prospects and even damaged China's reputation in general[25].

> **Box 2-2 Beyond DFIs: Environmental Governance Systems in the United Nations**
>
> Through United Nations mechanisms, nations have developed parallel systems to the systems of governance established by the DFIs profiled here. In this context, the CBD has long been a global platform for efforts to raise and harmonize national standards. CBD guidance is highly compatible with the "green BRI" framework, in which it encourages countries to collaborate in information sharing and capacity building to develop their own standards and practices (CBD, 1992, Article 14).
>
> In 2006, the CBD established voluntary guidelines for biodiversity-inclusive environmental impact assessment, including substantial upstream attention to identifying potential areas of concern. The guidelines encourage parties to focus upstream effort—before projects are proposed—in developing biodiversity mapping resources, such as the ones developed in China's recent history of demarcation of conservation priority areas. Individual project proposals can then be screened to ensure that all likely risks will be adequately addressed in the assessment stage. Impact assessments should be conducted with full participation by all stakeholders, to the extent possible. After individual projects' impact assessment, accountability mechanisms should be established to monitor and manage those projects' risks and oversee any necessary mitigation. CBD has also called for harmonization of standards among biodiversity financing mechanisms. Includes standards to apply in all cases, including but not limited to: highlighting and prioritizing the intrinsic value of biodiversity and its role in local livelihoods, effective public participation by project stakeholders, the establishment of institutional frameworks to oversee safeguard implementation.
>
> The Global Environment Facility (GEF) has been another important source of guidance on environmental standards. The GEF does not finance projects independently but rather works through co-financing. As such, its standards can "crowd in" other lenders and enable a broader reach. The GEF has nine minimum

standards for projects, including assessment, accountability mechanisms, conservation practices and restrictions on land use and the involuntary resettlement of existing communities. The first minimum standard, on environmental and social assessment, management and monitoring, echoes CBD guidance in its requirement for project screening as early as possible to establish which risks—among those covered by this standard as well as the remaining eight—may apply to each project. The second standard requires the establishment of institutional mechanisms such as those described below, to address problems that may arise in an accountable and transparent fashion. While the scope of these safeguards represents a crucial element in the environmental management of international development finance, its scale is modest. GEF's current 4-year work cycle draws on $4.1 billion in pledged funding (GEF, 2018). That represents a tiny fraction of the development finance issued through major development finance institutions. For comparison, the World Bank has approved over $120 billion in projects over the last four years[26]. For this reason, the international section of this paper focuses on the largest DFIs, which are the traditional sources of infrastructure finance in developing countries, as a comparison point for BRI projects.

Box 2-3 Beyond DFIs: Environmental Governance Systems in the Private Sector

In addition to the multilateral approaches profiled in this chapter, systems for private investment and finance have also made significant advances in recent years. Perhaps best-known are the *Equator Principles*, for use by private financial institutions in evaluating proposals for support. *Equator Principles* begin with an emphasis on early review and categorization of projects, to ensure that project-level assessments adequately address all of the salient environmental and social risks, in a way that ensures the broadest possible public participation. They also include the importance of well-designed institutional accountability mechanisms, which work in conjunction with national judiciary remedies to

> ensure appropriate project management in practice (*Equator Principles*, 2020). Complementary to the *Equator Principles* are the International Organization of Standards' environmental management tools, collected under the title ISO 14000. These systems do not specify specific safeguards but cover the extent to which institutions have established their own standards, with a commitment to employee training and auditing to ensure compliance.
>
> While these frameworks can be important tools for private lenders and investors to better select and manage projects, they are not strictly analogous to projects financed under the BRI, which involve cooperation among national governments. Thus, this section focuses on common practices among development finance institutions, which have traditionally represented the bulk of infrastructure finance for developing countries.

2.2.2 Comparative Analysis of Biodiversity Policy for International Financial Institutions

This section of the report surveys the practices of eleven major international institutions financing infrastructure across the world with respect to biodiversity. What immediately emerges from such an analysis is a remarkable convergence with respect to the objectives and guiding principles across these institutions. Virtually all institutions seek to minimize the risks to biodiversity and aim to have no net loss or even a net gain in biodiversity. Moreover, most institutions also require biodiversity assessments tied to mitigation measures, and entail stakeholder engagement and consultation in the assessment and management of biodiversity. A detailed analysis of specific operations and policies also shows that there are major similarities across institutions as well.

The majority of the international financial institutions have established the goal of biodiversity as a core of their activities. The Asian Infrastructure Investment Bank (AIIB), the Development Bank of Latin America (CAF), as well as the World Bank (WB) and International Finance Corporation (IFC) all recognize the need to "integrate conservation needs and development priorities; through sustainable use of the multiple economic, social and cultural values of biodiversity and natural resources in an optimized manner." To measure and calibrate such goals, institutions range from a policy of "no net loss" of biodiversity (such as the AIIB) or alternatively "no net loss or a net gain in biodiversity" [such as the European Investment Bank, EIB, Asian Development Bank (ADB), German Development Bank, and the CAF].

The majority of the international financial institutions also converge significantly with respect to overarching principles and policy operations for biodiversity protection. Virtually all of the institutions require these five traits:

- Alignment with international commitments and national legal requirements;
- Exclusionary lists of categorically ineligible projects due to biodiversity requirements.
- Requirements for biodiversity assessments and impact assessments;
- Application of a subsequent mitigation hierarchy for no net loss or a net gain to biodiversity;
- Meaningful stakeholder engagement and consultation in the assessment and management of biodiversity.

These policies are exhibited in Table 2-2 and Table A2-2 (in Annex 2). In Table 2-3, international institutions are listed vertically and specific biodiversity measures are listed horizontally across the table. It should be noted however that while these institutions have these policies, they are not always executed, which can thus lead to negative outcomes for projects, biodiversity and communities alike[27].

Table 2-2 Operational Requirements for Biodiversity Safeguards Applied by DFIs

	International Best Practice for Biodiversity Conservation				
	Alignment with International and National Commitments	Exclusionary list of categorically ineligible projects	Biodiversity Impact Assessments	Adopt Mitigation Hierarchy	Stakeholder engagement and consultation
ADB	X	X	X		X
AFDB	X	X	X	X	X

	International Best Practice for Biodiversity Conservation				
	Alignment with International and National Commitments	Exclusionary list of categorically ineligible projects	Biodiversity Impact Assessments	Adopt Mitigation Hierarchy	Stakeholder engagement and consultation
AIIB	X	X	X	X	X
BNDES	X		X	X	
CAF	X	X	X		X
EBRD	X	X	X	X	X
EIB	X	X	X	X	X
IDB	X	X	X	X	X
IFC	X	X	X	X	X
KFW	X	X	X		X
WB	X	X	X	X	X

Note: ADB: Asian Development Bank; AFDB: African Development Bank; AIIB: Asian Infrastructure Investment Bank; BNDES: Brazilian Development Bank; CAF: Development Bank of Latin America; EBRD: European Bank for Reconstruction and Development; EIB: European Investment Bank; IDB: Inter-American Development Bank; IFC: International Finance Corporation; KfW: German Development Bank; WB: World Bank.
Source: Authors' analysis of official documents and interviews.

(1) Alignment with international commitments and national legal requirements

A common trait across all of the international institutions is to align the practices of the institution with specific global or national commitments and legal requirements: Most of the institutions surveyed have language such as the following from the AIIB: "The Bank will not knowingly finance Projects involving the following... The production of, or trade in, any product or activity deemed illegal under national laws or regulations of the country in which the Project is located, or international conventions and agreements, or subject to international phase out or bans"[28]. The AIIB and others then provide an illustrative list of the kinds of international and national commitments they mean to adhere to (discussed below in "Exclusionary lists").

(2) Exclusionary lists of categorically ineligible projects due to biodiversity requirements

Often linked to the alignment language, the AIIB and others then provide an illustrative list of the kinds of international and national commitments they mean to adhere to. In the case of the AIIB they list the following[28]:

- Trade in wildlife or production of, or trade in, wildlife products regulated under the *Convention on International Trade in Endangered Species of Wild Fauna and Flora* (CITES).
- Activities prohibited by legislation of the country in which the Project is located or by international conventions relating to the protection of biodiversity resources or cultural resources, such as, *Bonn Convention*, *Ramsar Convention*, *World Heritage Convention* and *Convention on Biological Diversity*.
- Commercial logging operations or the purchase of logging equipment for use in primary tropical moist forests or old-growth forests.
- Production or trade in wood or other forestry products other than from sustainably managed forests.
- Marine and coastal fishing practices, such as large-scale pelagic drift net fishing and fine mesh net fishing, harmful to vulnerable and protected species in large numbers and damaging to marine biodiversity and habitats.

Most of the international institutions in the survey extend the possibility of excluding a project beyond these international and national commitments to cases where screening and environmental impact assessments may warrant it. Most have similar language on this matter: Such as the African Development Bank's policy reads that "If the Bank finds that the environmental or social impacts of any of its investments are not likely to be adequately addressed, the Bank may choose not to proceed with the investment...When the habitat/biodiversity implications of a project would appear to be particularly severe, the Bank may decide not to finance the project"[29].

(3) Requirements for biodiversity assessments and impact assessments

All of the major international institutions surveyed also perform relevant analyses of biodiversity impacts as part of broader environmental impact assessments. With respect to biodiversity, these policies charge the institution to consider the direct, indirect and cumulative project-related impacts on the habitats and the biodiversity they support. The World Bank considers threats to biodiversity, for example habitat loss, degradation and fragmentation, invasive alien species, overexploitation, hydrological changes, nutrient loading, pollution and incidental take, as well as projected climate change impacts. The World Bank also determines the significance of biodiversity or habitats based on their vulnerability and irreplaceability at the global, regional or national levels and will also take into account the differing values attached to biodiversity and habitats by project-affected parties and other interested parties[30]. Similar or identical language is found in the policies of most of the institutions studied here (Table 2-2).

The CAF is one institution with a slightly different language and scope. Its policy states that it will examine "Relevant physical, biological and socioeconomic conditions within the study area, In particular, environment-related aspects likely to be significantly affected by the proposed development, including, in particular, population, fauna, flora, soil, water, air, climatic factors, material assets, including the architectural and archaeological heritage, landscape and the interrelationship between factors above. Current and proposed development activities within the project's area of influence, including those not directly connected to the project"[31].

The Inter-American Development Bank (IDB) operates in such biodiverse places such as the Amazon basin which is home to many nations. The IDB's policy also addresses transboundary biodiversity issues associated with a project. The environmental assessment process for the IDB seeks to identify, early in the project cycle, transboundary issues associated with the operation. The environmental assessment process for operations with potentially significant transboundary environmental and associated social impacts, such as operations affecting another country's use of waterways, watersheds, coastal marine resources, biological corridors, regional air sheds and aquifers, will address the following issues: (i) notification to the affected country or countries of the critical transboundary impacts; (ii) implementation of an appropriate framework for consultation of affected parties;

(iii) appropriate environmental mitigation and/or monitoring measures, to the bank's satisfaction.

In addition to estimating biodiversity impacts, international bodies recommend that economic impacts be differentiated on a gender basis, in order to estimate the indirect impact on women's work as stewards of crop biodiversity. The CBD's *2015-2020 Gender Action Plan* calls for calculating project costs and benefits should be estimated differently for women and men, rather than collectively, as do the Green Climate Fund and Climate Investment Funds[32-34].

(4) Application of a subsequent mitigation hierarchy for no net loss or a net gain to biodiversity

To the extent that the compulsory biodiversity impact assessments identify issues that may impact biodiversity, Table 2-2 shows that most of the major international financial institutions (8 of the 11 surveyed) require a mitigation hierarchy to meet the overall objective of no net loss or a net gain in biodiversity. The mitigation hierarchy has the following four pillars:

- **Avoidance**: measures taken to avoid creating impacts from the outset, such as careful spatial or temporal placement of elements of infrastructure, in order to completely avoid impacts on certain components of biodiversity.
- **Minimization**: measures taken to reduce the duration, intensity and/or extent of impacts (including direct, indirect and cumulative impacts, as appropriate) that cannot be completely avoided, as far as is practically feasible.
- **Rehabilitation/restoration**: measures taken to rehabilitate degraded ecosystems or restore cleared ecosystems following exposure to impacts that cannot be completely avoided and/ or minimized.
- **Compensation:** measures, such as offsets, taken to compensate for any residual significant, adverse impacts that cannot be avoided. Offsets can take the form of positive management interventions such as restoration of degraded habitat, arrested degradation or averted risk, protecting areas where there is imminent or projected loss of biodiversity.

(5) Meaningful stakeholder engagement and consultation in the assessment and management of biodiversity:

All of the institutions surveyed for this SPS require stakeholder engagement and consultation in the assessment and management of biodiversity. Each of the institutions makes some commitment to carry out consultations with affected peoples and communities and seek their informed participation throughout the project cycle.

As noted earlier, the CAF is perhaps the most engaged in major infrastructure projects in areas where there are significant concerns about biodiversity in areas that inhabit large and often vulnerable populations. The CAF requires that consultations with project-affected groups be held early in the environmental impact assessment process and maintained throughout the project cycle. Throughout the project cycle important information is supposed to be disclosed in a timely manner to affected groups, civil society organizations and other key stakeholders. The CAF also requires that "The potential impact of projects over forests and natural habitats, and the rights of access to and use of resources for the welfare of the communities shall be evaluated as a part to the Environmental and Social Assessment" (CAF,2015, 64).

The IFC requires that borrowers go so far as implementing a *Stakeholder Engagement Plan*. Where applicable, the *Stakeholder Engagement Plan* will include differentiated measures to allow the effective participation of those identified as disadvantaged or vulnerable. When the stakeholder engagement process depends substantially on community representatives, the client is required to make every reasonable effort to verify that such persons do in fact represent the views of affected communities and that they can be relied upon to faithfully communicate the results of consultations to their constituents. When affected communities are subject to identified risks and adverse impacts from a project, the client will undertake a process of consultation in a manner that provides the affected communities with opportunities to express their views on project risks, impacts and mitigation measures and allows the client to consider and respond to them[35].

Development finance institutions have learned the importance of ensuring that their stakeholder engagement plans incorporate the voices of women, particularly in cases where communities may be facing displacement. As mentioned in section 1.4.2 above, in many rural, poor settings around the world women do not customarily take part in public discussions but do bear the brunt of biodiversity losses, which can curtail their ability to serve as stewards of crop biodiversity, further potential biodiversity losses, and limit the benefits of conservation projects. For example, an inter-bank working group with representatives from the AIIB, ADB, AFDB, EBRD, EIB, IDB, NDB and the World Bank recently published joint recommendations on meaningful stakeholder engagement, which encourage project planners to ensure that these processes are designed specifically to prioritize the participation of women and other disadvantaged groups, and if necessary, disaggregate stakeholder engagement processes by gender[36].

2.3 Areas of Convergence and Divergence between China and International Peers

China has made significant progress toward biodiversity-related SDGs, including SDG 15. *China's Sixth National Report on implementing CBD* shows satisfactory progress in most areas, including several that are highly relevant to the BRI: integrating biodiversity into national and local planning, mobilizing and increasing biodiversity finance, and conservation of inland freshwater, forest and mountain ecosystems. In these three areas, significant synergies exist between China's areas of growth and those of its peers, and these areas of overlap can be fertile ground for increased cooperation in the context of the BRI.

One of the most important mechanisms China has used to accomplish its domestic progress has been through the use of the Ecological Conservation Red Line (ECRL) system, protecting ecologically fragile regions and those that provide crucial ecological services. With a framework established in early 2017 by the General Office of the CPC Central Committee and the General Office of State Council, ecological red lines have quickly been developed nationally. In 2018 the State Council of China approved the red line plans from 15 provinces (autonomous regions and municipalities), including Beijing, Tianjin, Hebei, Ningxia and the provinces and municipalities of the Yangtze River Economic Belt. By the end of 2020, the remaining national territory should be included in the red line system. The total protected land is expected to reach roughly-one third of the entire landmass of China.

Internationally, biodiversity conservation progress has centered around the establishment and mainstreaming of global standards that establish not only geographic limits, but also operational risk management strategies to protect local ecosystems, institutional reputations and the cooperative relationships of all the partners involved in projects. Major international development finance institutions—including both multilateral and national development banks—use five main approaches to operationalizing international standards: ① aligning institutional practices with international or national commitments; ② using exclusionary lists of categorically ineligible projects; ③ requiring projects to undergo biodiversity impact assessments; ④ adopting a mitigation hierarchy to do no harm and, if possible, benefit local ecosystems; ⑤ incorporate local stakeholders.

These approaches don't need involve development lenders dictating terms to borrowing countries or single-handedly managing projects. Instead, they can induce international collaboration toward shared goals, built on information sharing from multiple nations' experience and accumulated expertise. For example, international bodies have developed platforms that can aid in cooperative approaches to the application of the mitigation hierarchy. The Restoration Opportunities Assessment Methodology (ROAM), developed by the IUCN and WRI is one such mechanism. ROAM supports national governments' development of ecosystem restoration programs, including the identification of priority areas, cost-benefit analyses for intervention approaches and financing options[37]. In this regard, the green BRI is an apt platform for furthering international cooperation to prioritize biodiversity conservation in development planning, showcasing China's experiences and building capacity among BRI participating nations.

Among these five approaches, China's ECRL system is most closely aligned with the use of a mitigation hierarchy, which emphasizes the first preference for avoidance of harm, by respecting areas associated with ecological fragility or important ecological services. The strong synergy of these techniques makes China's ECRL approach highly relevant for incorporation into BRI planning. Other BRI participating countries are familiar with the approach and China has found great success with it domestically. Incorporating ECRL planning—in the context of international collaboration—into BRI project planning could be a powerful tool in aligning the BRI with SDG 15. In March 2020, SDSN Executive Director Guido Schmidt-Traub [38], writing for China Dialogue-Learning Ecological Redline from China, notes that "China is the only country practising such comprehensive and ambitious land-use planning" and that "China's experience may be relevant for any country wishing to meet the objectives of the CBD and the Paris Agreement". According to Schimidt-Traub, the inclusion of land-use maps in climate and biodiversity strategies would enable the success of CBD COP 15 and UNFCCC COP 26, while land-using planning itself can serve as a critical tool for directing the economic stimuli in the right direction after the COVID-19 pandemic.

3 ANALYSIS OF SDG 15 RELATED INVESTMENTS POSSIBILITIES

Working towards SDG 15 is no small task. Biodiversity is fragile and necessary for the lives and livelihoods of global communities, and if damaged, it would be difficult or impossible to regenerate. To prioritize it among the ever-accelerating world of international finance and investment, the field of biodiversity finance has emerged.

The need is certainly present and pressing. At a 2015 workshop in Beijing, the Intergovernmental Science-Policy Platform on Biodiversity and Ecosystem Services (IPBES) concluded that "urgent and concerted action" was needed to avert ecosystem degradation globally, for the sake of 3.2 billion people currently impacted by degraded lands[39]. Economically, they estimated that the losses caused by this biodiversity degradation amount to 10% of global GDP. These same authors track successful ecosystem restoration across every region and continent of the globe.

Biodiversity conservation is, by definition, an act that prioritizes long-term well-being over short-term booms. It requires investing in the natural capital necessary to support future economic production and human health. It also requires investing in activities that will pay off in positive externalities distributed throughout a wide array of communities, which the investor will not be able to completely reap themselves. Thus, it needs external encouragement in order to flourish, in the form of an enabling policy environment, preferential financial arrangements, and impact investors motivated to fuel positive change not only for their own portfolios but for the communities where they operate.

3.1 Survey and Assessment of the Chinese Experience

The SDG 15 aims to protect, restore and promote the sustainable use of terrestrial ecosystems. In recent years, China has continuously increased its financial input in ecological compensation mechanisms, transfer payments to ecological function areas, grassland compensation, subsidies for returning farmlands to forests, subsidies for wetland protection and restoration and other programs. In the meantime, China has continued improving the property right system of natural resources, exploring new ways of cooperation among governments, businesses and environmental organizations, promoting sustainable forest management, combating desertification, halting and reversing land degradation, and halting biodiversity loss. In 2018, China scored 62.7 on SDG 15, up by 7% compared with 2017, indicating that certain progress has been achieved in terms of terrestrial ecosystem protection.

The ecological compensation mechanism continues to improve[40]. The Chinese government attaches great importance to the development of the ecological compensation mechanism and launched policy documents such as *Suggestions on Improving the Ecological Compensation Mechanism*; *Guidelines for Accelerating the Development of a Horizontal Ecological Compensation Mechanism for Upper and Lower Reaches of Rivers*; *Action Plan for the Establishment of a Market-Oriented, Diversified Ecological Compensation Mechanism*; *Guidelines for Establishing and Improving the Long-term Mechanism for Ecological Compensation and Conservation in the Yangtze River Economic Belt*; and *Plan for A Pilot Program of Establishing a Comprehensive Ecological Compensation Mechanism*. These documents establish the framework of an ecological compensation mechanism with Chinese characteristics. China had a fiscal input of nearly RMB 200 billion yuan in ecological compensation in 2019. Meanwhile, both the central and local governments have been taking market-oriented approaches to expand the source of funding for improving the ecological compensation mechanism. For example, the water source areas of the middle route of the South-to-North Water Diversion Project established ecological compensation through pairing cooperation; Jinhua City and Pan'an City in Zhejiang Province took the lead to adopt off-site development as a means of compensation; the drainage areas of Xin'an River engaged the private sector in ecological compensation programs; Moutai Group plans to invest a total of RMB 500 million yuan in 10 years starting from 2014 in water environment compensation in the drainage areas of Chishui River; and China Three Gorges Corporation has been playing an active role in the protection of the Yangtze River while exploring for market-oriented approaches to improve the compensation mechanism.

Transfer payments to ecological function areas have been increasing. To guide local governments to intensify the efforts to protect the ecological environment and improve the capacity of local governments in places with national key ecological function areas to provide basic public services, the central government established the transfer payment system for key national ecological function areas in 2018 to support the protection of these areas. By the end of 2019, the central government has made transfers amounting to RMB 524.2 billion yuan to key national ecological function areas, of which RMB 81.1 billion yuan was made in 2019, RMB 9 billion yuan more than in the previous year, registering an increase of 12.5%. Meanwhile, China has kept expanding the coverage of key national ecological function areas to 819 counties. Once included in national ecological function areas, local governments will receive financial and policy support as long as they strictly implement the negative list system for industrial access. According to relevant regulations, a region counted as a key national ecological function area needs to strengthen ecological protection and restoration, regulate the boundaries of industrialization and urbanization and enhance the supply capacity of eco-products.

The standards for the compensation for ecological services of forests have been rising. In recent years, the central government has been increasing its input in compensation for ecological services of forests and raising the standards for compensation year by year. Starting from 2010, the standards of compensation for state-level non-commercial forests have varied according to their ownership. The compensation rate for state-owned state-level non-commercial forests was RMB 5 yuan/(year·mu) (1 mu is about 0.067 hm^2) in 2010, while that of privately-owned and community-owned state-level non-commercial forests has increased from RMB 5 yuan/(a·mu) to RMB 10 yuan/(a·mu). In 2013, the compensation rate for privately-owned and community-owned state-level non-commercial forests was raised to RMB 15 yuan/(a·mu). In 2015, 2016 and 2017, the rate for state-owned state-level non-commercial forests increased step by step, reaching RMB 6 yuan/(a·mu), RMB 8 yuan/(a·mu) and RMB 10 yuan/(a·mu) respectively. As the central government increases its fiscal input and raises the standards for compensation, local governments are expected to positively improve compensation system for ecological services of forests in local areas.

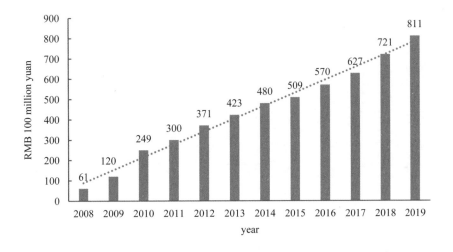

Figure 3-1 Transfer Payment to Key National Ecological Function Areas (2008-2018) [41]

Policies for fiscal support to ecological protection and restoration of wetlands continue to improve. China attaches great importance to the protection of wetlands with increasing fiscal input in accelerating the development and optimization of policies concerning fiscal support to ecological protection and restoration. From 2013 to 2016, the central government allocated RMB 5 billion yuan to protect wetlands in China and continued to provide support through the Funds for Reform and Development of Forestry afterwards. In 2014, the Ministry of Finance and the State Forestry Administration launched the pilot program of wetland ecological benefit compensation. For important wetlands on the route of migratory birds managed by the forestry system, their loss due to the protection of birds and other wild animals will be properly compensated. Currently, the central government allocates fiscal input to local governments, who will then decide the scope of wetland ecological benefit compensation and the areas to be protected.

> **Box 3-1 Measures Taken by the Funds for Reform and Development of Forestry to Support Wetland Protection and Restoration**
>
> The first measure entails supporting the protection and restoration of wetlands. For wetlands of international/national importance, national wetland parks at important ecological locations, and national wetland nature reserves at or above the provincial level managed by the forestry system, efforts will be made to protect and restore the wetlands, improve the current ecological status, and maintain the health of the local ecosystem.
>
> The second measure entails supporting the restoration of farmland to wetland. It is encouraged to return farmlands to wetlands within the wetlands of international importance, national wetland nature reserves and provincial nature reserves within wetlands of national importance managed by the forestry system, so as to expand the area of wetlands and improve the surrounding ecological status.
>
> The third measure entails supporting the wetland ecological benefit compensation. For important wetlands on the route of migratory birds managed by the forestry system, their loss due to the protection of birds and other wild animals will be properly compensated. In so doing, all parties are motivated to protect wetlands and maintain the wetlands' ecosystem service functions.

The grassland ecological protection subsidy incentive policy has been continuously promoted. To protect grassland ecosystem, guarantee the supply of meat and dairy products and increase the income of herders, the Chinese government implemented the grassland ecological protection subsidy incentive policy. Currently, it covers eight major pastoral provinces (autonomous regions), including Inner Mongolia, Xinjiang, Tibet, Qinghai, Sichuan, Gansu, Ningxia and Yunnan, and five non-major pastoral provinces, such as Heilongjiang. RMB 152.033 billion yuan has been given as subsidies to 268 pastoral and farming-pastoral counties in the above provinces. In 2019, a new round of grassland ecological protection subsidy incentive of RMB 18.76 billion yuan was included in the central budget to support the banned grazing area of 1.206 billion mu and the grass-animal balance area of 2.605 billion mu.

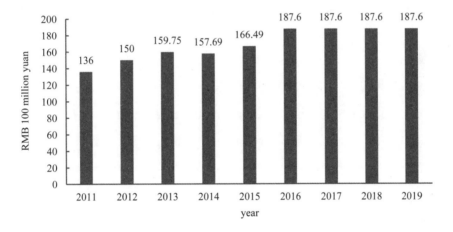

Figure 3-2 Grassland Ecological Protection Subsidy Incentive (2011-2019)

Source: Chinese Academy of Environmental Planning (2018) and Ministry of Finance website.

The unified confirmation and registration of natural resources has been rolled out. The confirmation and registration of natural resources is important in promoting the reform of the property rights mechanism of natural resource assets, which is a key part of China's ecological civilization framework. By the end of October 2018, 1,191 natural resource registration units have been established in 12 provinces (autonomous regions) and 32 pilot areas, and the total registered area has reached 186,727 km^2. The state also focused on exploring the confirmation and registration of national parks, wetlands, water flows and proven reserves of mineral resources. Starting from the end of 2018, the confirmation and registration of natural resources in key areas has been implemented nationwide step by step. It is planned that within 5 years, the unified confirmation and registration of natural resources in nature reserves will be completed, such as national and provincial key parks, natural reserves and various natural parks (scenic areas, wetland parks, natural heritage, geo-parks, etc.). At the same time, the unified confirmation and registration of individual natural resources with complete ecological functions owned by the public will be conducted, such as major rivers and lakes, key wetlands, key national forests, important grasslands and other areas.

Innovative approaches have been taken to promote cooperation between governments, environmental organizations and large enterprises. "Debt-for-nature swaps" can be traced back to the 1980s. In a debt for nature swap, a nation agrees to swap the preservation of the natural environment for some of its debt. This benefits the nation because it brings its overall debt level down, and it benefits the environment by creating more protected habitat for animals and plants. "Debt-for-nature swaps" may be organized by conservation organizations or by governmental organizations concerned with environmental preservation. Currently, China doesn't have any recorded cases of debt-for-nature swaps. However, there are many cases in which international organizations or large enterprises cooperate with governments in environmental protection for win-win benefits, mostly in the form of Payments for Ecosystem Services (PES). The project for the protection of water source jointly operated by The Nature Conservancy (TNC) and Longwu, Zhejiang Province is a case of such a practice. with TNC as the consultant, The project, protects water source areas with funding from trust agencies. The forest land is operated and managed in an integrated manner. The income from such operations is used to cover the compensation to farmers and the cost of the protection and management of the water source area. In this way, the water source could be properly protected and the trust fund company could get a fair share of economic return.

Box 3-2 Water Source Protection Program Jointly Operated by TNC and Longwu, Zhejiang Province in the Form of Trust [42-43]

On January 15, 2015, TNC and the government of Huanghu County, Zhejiang Province, signed an agreement on the protection of Longwu Reservoir, a county-level water source. According to the agreement, the goal of the program is to reduce the factors that may cause water quality deterioration and improve the water quality of the reservoir from Class II to Class I. It is a good attempt to adopt a win-win model towards ecological conservation that could benefit both the environment and the community. Funded by Alibaba Foundation, it is the first water source protection program of TNC in China in the form of trust. In September 2015, Wanxiang Trust-TNC Charitable Trust decided to invest RMB 330,000 yuan to support the program.

In November 2015, Wanxiang Trust launched the first water fund trust in China—Wanxiang Trust-Shanshui Fund No.1. Shanshui Fund Trust invited TNC to be the consultant. In the same month, the first water source protection and management project supported by the Trust—Longwu Small Water Source Protection Program was officially launched.

The program is an innovation that integrates social resources and engages multiple actors, including farmers, financial institutions, charity organizations, local organizations, businesses in the lower stream of the industrial chain related to agriculture in its daily operations. In this way, it could generate positive outcomes through interaction, collaboration and sharing. The program effectively addressed the issue of pollution caused by human activities that could benefit the whole community; it also established a sustainable funding mechanism that could bring environmental benefits and economic returns for investors.

Mode: Trust;

Target of compensation: Residents in the nearby community;

Main provider of compensation: Alibaba Foundation;

Means of compensation: Farmers entrust the forest land to the trust and get steady income as compensation;

> Operation mechanism: The trust agency operates and manages the forest land while promoting the growing of bamboo shoot and ecological tourism. The income from such operations is used to cover the compensation to farmers and the cost of the protection and management of water source area.

3.2 Survey and Assessment of International Practices

Globally, biodiversity finance has taken a variety of approaches, often in the form of PES or investments in soil and water health to ensure their ability to support future agriculture. The former category has been explored among developed as well as developing countries, under a variety of terms including "eco-compensation" in China and "agri-environmental programs" in the European Union[44]. A recent global survey in the journal *Nature Sustainability* estimates that over $36 billion is invested in PES annually, with approximately one-third of that amount taking place within China[45]. Of the conservative USD 36 billion estimate, the great majority is estimated to be concentrated in watershed subsidies, which attract approximately USD 23.7 billion per year. One of the better-known examples of watershed PES programs among developing countries is found in Ecuador, where the capital city of Quito established the world's first municipal water fund, with the cooperation of TNC, in 2000 (Echavarría, 2002). Quito's groundbreaking PES program (named FONDAG for the Spanish acronym for Water Protection Fund) added a surcharge on water users and included a bottled water plant in order to fund the conservation of the watershed that provides water for the municipality of Quito.

A second common approach—and one of particular importance for the context of a green BRI—is the use of biodiversity offsets. These financial arrangements seek to mitigate the *net* biodiversity impact (or create a positive net biodiversity impact if possible) due to new project construction by financing separate conservation efforts. Enabling policies for this type of biodiversity financing exist throughout Asia, Europe and the Americas, but range widely in definition[46-48]. A recent study in *Nature Sustainability* which considered only those projects implemented under a no net loss policy, found nearly 13,000 such projects globally, cover an area of roughly 153,679 km^2. Some of the largest existing such programs have occurred in very disparate countries, including: Mongolia, Brazil and Uzbekistan[49]. However, as Gardener et al. (2013)[50] demonstrate in a *Conservation Biology* article, the no net loss standard is highly ambitious in practice, requiring comparable gains of new biodiversity (not simply conservation) relative to the biodiversity losses that are to be offset, and requiring that those gains be maintained over the long term. Accomplishing these goals requires strong institutional support and the involvement of a wide array of geographic locations, in order to effectively "hedge" the risks of partial project failure.

More broadly, biodiversity offsets take place within the mitigation hierarchy described in section 2.2.2. As part of the "compensation" stage, offsets serve as a last option if the earlier options of avoiding, minimizing and rehabilitating/restoring ecosystems and the communities that rely on them are not feasible or insufficient. For example, Villarroyo, Barros and Kiesecker (2014) reviewed the national policies regarding biodiversity offsets in Latin America and found that three countries' national regulations (Chile, Colombia, and Mexico) specifically made mention of both the mitigation hierarchy and offsets in relation to their EIA processes. However, many scholars note that significant institutional capacity building work remains in governments that seek to support offset schemes, particularly in establishing the scientific bases for "ecological equivalence" between geographic areas, in order to be assured of the net biodiversity impact of offset schemes[49-52].

Nonetheless, Luck, Chan and Fay (2009)[53] found that global biodiversity finance had suffered from a severe geographical mismatch: flows had been directed mostly to low-priority ecosystems, while the most important ecosystems had been overlooked. Over half of all flows were focused on the United States, Canada, and Europe, despite the fact that these ecosystems were of mutual low priority for the two goals of preserving ecosystem services and biodiversity. Meanwhile, areas of high priority for both of those goals were concentrated in Southeast Asia and South America, which had attracted less than 15% of global biodiversity collectively. This mismatch is assuredly related to the fact that approximately half of all biodiversity finance stems from domestic government funding, and so funding is staying within wealthy countries. Thus, if the world is to make progress toward SDG 15, international biodiversity cooperation—through investment and especially aid—will be key.

Incorporating Commercial Investors

Traditionally, biodiversity finance has been limited to aid: official development aid (ODA) and philanthropy. However, opportunities for commercial investors have also been developing in recent years. Many types of biodiversity-maintaining or biodiversity-enhancing activities will pay for themselves in the medium-to-long term, though they require new sources of upfront financing to launch. By preserving or enhancing existing natural capital, these investment possibilities allow for reduced costs in economic production in the long term. For example, Burian et al. (2018)[54] advocate for agricultural investments aimed at building soil health and resilience, which will bring economic benefits in increased crop yields and decreased expenses on agrichemical inputs. IPBES estimates that the economic benefits of soil preservation are an average of 10 times greater than the cost of those efforts[39]. Finally, these benefits are multiplied as they impact downstream ecosystems through less-polluted waterways that better support both urban and rural life.

However, to succeed, biodiversity finance efforts must be well-matched with local needs, well-designed with local input and well-managed by local governments. Clark, Reed and Sunderland (2018)[55] find that the sector is beset by potential "greenwashing," in which commercial investors' activities are not actually biodiversity—enhancing or biodiversity—protecting, but market themselves as such in order to access advantageous financing and public reputational enhancements. While a few such investments may do no harm, allowing this type of activity to flourish under the banner of "biodiversity finance" brings risks to the entire sector, threatening the legitimacy of its claims and with it, its access to the favorable financing that will ensure its continued viability.

Bearing in mind the potential benefits as well as the potential risks, the United Nations Development Programme (UNDP)'s "Biodiversity Finance Initiative" (BIOFIN) has developed five areas of focus for developing frameworks for commercial biodiversity finance.

- Policy and Institutional Review, examining the ways in which national institutions are sufficiency robust and well-designed to encourage biodiversity finance, as well as what areas can benefit from reforms (with the added optional aspect of identifying economic drivers of biodiversity loss);
- Expenditure Review, calculating current expenditures to support biodiversity;
- Needs Assessment, estimating the total amount necessary in biodiversity-supporting expenditures and the gap in actual spending;
- Finance Plan, setting targets and finding potential sources for this funding;
- Finance Solutions, establishing and carrying out a plan to address the institutional and financial gaps discovered in previous steps.

Bilateral BIOFIN Cooperation Between BRI participating countries

As the sector of biodiversity finance continues to expand, and particularly as it opens to commercial activities, China has the opportunity to establish itself as a global leader in the field. The globally networked nature of the BRI highlights the importance of working to preserve biodiversity in the "hotspots" along the network to ensure that the entire enterprise brings net benefits to the communities and the ecosystems that support them.

Two such "hotspots" stand out among potential partners for Chinese conservation finance, one on each side of the Pacific Ocean: Indonesia and Ecuador—two countries with BRI MOUs with China. Both are among the 17 mega-biodiverse countries, who collectively boast 17% of the world's species. They represent the peak of global terrestrial and marine biodiversity, respectively. Ecuador is home to the most biodiverse section of the Amazon rainforest, which is often called the "lungs of the earth" [56]. The Ecuadorian Amazon rainforest sits at the headwaters of the Amazon River, and ecosystem preservation there has the potential to benefit the downstream Amazon biome. Indonesia sits in the center of the world's marine biodiversity, sometimes called the "Amazon of the Seas" or the Coral Triangle, for the tremendous density and diversity of coral species there (Hoeksema, 2007). Both countries have made significant progress in their UNDP BIOFIN process, preparing to host and manage biodiversity finance successfully.

Furthermore, both Indonesia and Ecuador have strong economic ties with China. According to FDI Markets, China has created more new investment in Indonesia than in any other country in the last decade: over USD 52 billion. Ecuador, while a much smaller country, has also built up an important friendship with China. For the last decade, China has been Ecuador's most important creditor, creating deep goodwill between the two governments. In 2019, Ecuador became the first Latin American or Caribbean nation to become a full member of the AIIB, signaling its interest and institutional readiness to continue to strengthen its financial ties with Asia and in particular, with China.

One major obstacle to biodiversity preservation is a simple matter of geography: biodiversity hotspots are disproportionately located in developing countries, with limited fiscal space to design and carry out long-term projects that will not yield financial benefits for many years. One way to circumvent this problem is for these countries to collaborate in a bilateral or plurilateral fashion with their most important strategic partners, either creditor nations or major sources of investment, to ensure that the growth in economic activity between their nations does not bring environmental degradation. Three main models are common in the area of bilateral biodiversity finance: debt-for-nature swaps, National Environmental Funds (NEFs) and bilateral sustainable development banks.

In debt-for-nature swaps, creditors cancel a share of the debt in exchange for the debt service payments being redirected to maintain biodiversity. Alternatively, impact investors or international non-profit organizations play a pivotal role, negotiating a deal in which they buy a country's debt at a discount, work with the debtor nation to build the institutional infrastructure to oversee the biodiversity plan and help establish a fund to support these activities. These deals can effectively cut off a vicious cycle of fiscal constraints leading to insufficient environmental management, hurting economic outcomes, reinforcing fiscal constraints.

When implemented well, debt-for-nature swaps can allow chronically indebted countries an alternative to environmentally-damaging activities to pay down debt. They can also create an institutional structure to oversee the establishment of definitions of sustainable economic activities appropriate for the newly protected areas, and the fiscal space to ensure that the new protections are well-managed, with adequate participation from local communities to ensure enforcement. However, debt-for-nature swaps are not quick fixes for serious debt problems, nor can they bring a sudden stop to ongoing ecological disasters. As the case of Seychelles demonstrates, establishing the conservation areas is a process of multiple years. Thus, rather than being used as a last resort or rescue option for disaster scenarios, it is best considered as a long-term, proactive approach to conservation.

NEFs share many of the same characteristics of "debt-for-nature swaps", but with less intervention on the part of outside partners. NEFs are locally-managed funds set up in collaboration with external partners that support conservation efforts domestically. The "trust fund" nature of NEFs can make them particularly appropriate funding instruments for projects that need medium- or long-term investments, such as the delineation, establishment and maintenance of national parks. For example, Brazil's Amazon Fund supports non-deforestation livelihood projects for forest-dwelling communities[57]. Other Amazon-basin countries including Bolivia, Colombia and Peru have NEFs to support their national systems of protected areas. In Asia, Bhutan and the Philippines both have similar funds[58].

While NEFs are managed by national governments, they can be established in conjunction with strategic partners overseas. For example, the Foundation for the Philippine Environment has been supported through debt swaps from the United States and Japan. In these cases, NEFs are similar to the "debt-for-nature swaps" profiled above, without the same level of detailed conditionality. Instead of debtor nations agreeing to set aside particular tracts of land, they establish general support for the nationally-defined conservation strategies. The fact that the local governments oversee the funds and their management makes them suitable for bilateral cooperation with partners that prefer to allow as much local direction as possible.

Finally, bilateral conservation financing can take the form of special-purpose development banks. For example, the North American Development Bank is a project of the governments of the United States and Mexico and was brought into being as part of the negotiations for the North American Free Trade Agreement (NAFTA), with the objective of ensuring that the U.S.-Mexico border would not be degraded due to the higher economic activity expected under NAFTA. It finances sustainable development projects on both sides of the border[60]. It has financed USD 1.2 billion in projects as of year-end 2018 (North American Development Bank, 2019). This model may be particularly attractive in the establishment of cross-border transit corridors such as those in the BRI, or in partnerships between countries that expect to see significant increases in investment and trade.

Box 3-3　"Debt-for-Nature Swap" in the Seychelles

Nature Vest, the biodiversity finance platform of TNC was founded in 2014 to mobilize private capital for conservation. In 2016, in conjunction with other private funders, Nature Vest signed a deal with the Seychelles' Paris Club creditors to buy a portion of Seychelles' debt at a discounted rate, spending approximately $22 million for approximately $25 million in debt.

In cooperation with the government of Seychelles, this debt relief will allow the establishment and maintenance of approximately 400,000 km² of ocean. As of this writing, roughly half of that area has already been set aside, in the form of two new protected areas. The remainder is expected to be added within a year of this writing.

Two factors have contributed to the success of this "debt-for-nature swap": the leadership of the government of Seychelles and the unhurried nature of the planning process. Through both of these strategies, this Seychelles project has been able to earn the local support necessary for management and enforcement over the years to come.

This process represents a culmination of existing national government goals announced in 2012, when it announced a plan to increase protected areas to include 30% of its marine EEZ . Seychelles adopted a mapping methodology using international best practices, adapting UNESCO[61] recommendations, i.e. The Seychelles Marine Spatial Planning (MSP) initiative, beginning in 2014 ("Seychelles Marine," n.d.). The MSP has been a deliberately time-consuming process in order to ensure an evidence-based outcome with sufficient public input. In fact, while Phase I was completed in 2018, protecting 15% of the Exclusive Economic Zone (EEZ), Phase II is not expected to finish until the end of 2020. To determine which marine areas would be protected and what sustainable activities would be permitted, the MSP incorporated the input of 10 ministries and 100 public stakeholders who participated through 9 public workshops and 60 consultations.

3.3 The Applicability of Chinese Domestic Experience for Biodiversity Finance with International Peers

Both within China and globally, biodiversity finance has grown tremendously in the last decade. PES programs have been popular internationally for many years but have skyrocketed in the last decade within China. China's peers have also more deeply explored other types of biodiversity finance, such as offsets and debt-for-nature swaps. In these areas, China's scientific expertise could help BRI participating countries face some of those avenues' obstacles, and China can also learn from international successes.

While many countries have pursued PES programs, the depth, scope and rapidity with which China has developed its internal programs hold important lessons for its peers. Programs to support reforestation (differentiated between purely ecological forests and commercial forests), farmland conversion to grasslands (differentiating between no-graze areas and mixed areas) and wetlands (including special compensation for lands along important migratory bird routes) show a level of scientific expertise and local understanding that many other BRI national governments could draw reference in accordance with their own conditions. The collaborating context of the BRI participating countries allows for information sharing and collaborative planning. In this context, it would be wise to share these experiences where possible.

Globally, biodiversity finance has been a mix of administrative and market mechanisms. In some instances, governments manage programs entirely, especially those focused on payments for environmental services. However, in the cases of offsets, where biodiversity enhancements are specifically linked to biodiversity loss caused by commercial investment, governments have focused on establishing legal frameworks to enable biodiversity offset mechanisms. While biodiversity offsets have grown popular in developed and developing countries alike, the complexity of managing multiple geographic sites (where biodiversity is being lost and gained) creates significant demand for institutional capacity. Specifically, scientific expertise is needed to establish an ecological survey to accurately establish equivalence between sites and measure outcomes. Even though China does not specifically link biodiversity—enhancing programs with biodiversity—damaging building projects through offsets, it has had to develop this same type of institutional capacity to establish its ECRL system. Situations where BRI participating countries are pursuing offsets may be excellent opportunities for collaboration in this area.

A 2019 joint report between the China Development Bank and the UNDP identified key benefits for collaboration and harmonization of standards and across the BRI participating countries and religions, summarized in Table 3-1. From this research, it is clear that all partners will be made better off by harmonizing standards and sharing expertise from China's ECRL system, partners' biodiversity finance experience and shared priorities for conservation.

Table 3-1 Benefits of Harmonizing Standards on BRI participating countries

Nation	Beneficiary	Benefit
China	Government	● Prevention and mitigation of reputational risk; ● Prevention and mitigation of environmental and social risks.
	Financial institutions	● Prevention of default risk; ● More innovative and competitive financial services.
	Business entities	● Expansion of markets; ● Greater returns and effectiveness of projects; ● Improved competitiveness and risk management.
Partners	Governments	● Economic growth and poverty alleviation; ● Improved management of fiscal resources; ● Improved management of natural resources; ● Strengthening of institutional capacities; ● Prevention and mitigation of environmental and social risks; ● Reduction of compliance costs.
	Business entities	● Expansion of domestic markets and linkage to international value chains; ● Increased opportunities for participation in BRI-related procurement; ● Improved compliance and risk management.
All	Local communities	● Improved livelihoods, decent jobs; ● Reduced likelihood of social conflict; ● Enhanced voice and ownership of projects; ● Reduced vulnerability against potential negative impacts.
	Global community	● More equitable use of resources and growth; ● Enhanced interconnectivity and cooperation; ● Provision of global public goods; ● Improved global governance.

Source: Reproduced from UNPD and CDB (2019).

4 ANALYSIS OF SDG15-RELATED GOVERNANCE STRUCTURE

4.1 Survey and Assessment of Practices in China

4.1.1 China's Biodiversity Conservation and Governance Framework

China has integrated biodiversity into the target system for building an Ecological Civilization, and has been constantly improving the system and institutional settings for biodiversity conservation. It employs a system characterized by unified national regulation and division of responsibilities and cooperation among different sectors to protect biodiversity. In particular, right after it approved the CBD in 1993, China established a Coordinating Group for the Implementation of CBD, with the former State Environmental Protection Administration (SEPA) as the leading agency and the participation of 20 departments/line ministries under the Chinese State Council. It founded a CBD implementation office in the then SEPA and identified the national focal points for CBD implementation, biodiversity clearing-house mechanism, and biosafety respectively. An inter-ministerial Joint Meeting for Protection of Biological Resources was set up at the same time. The Coordinating Group meets every year to develop an annual work plan for CBD implementation and launches a variety of activities. So far, an initial national working mechanism has been formed for biodiversity conservation and CBD implementation. China started the development of a China Biodiversity Conservation Action Plan in 1992 and released the finalized document in 1994. Referred to as "Action Plan", this Action Plan has identified both the location of ecosystems and the list of species for priority conservation, and set out the goals for seven areas of biodiversity conservation in China.

In 2010, the State Council of China founded the "China National Committee for the 2020 International Year of Biodiversity". During its meeting, the State Council reviewed and approved the *China Action Plan for the International Year of Biodiversity* and *the China National Biodiversity Conservation Strategy and Action Plan (2011 - 2030)*. In June 2011, the State Council decided to rename the "China National Committee for the International Year of Biodiversity 2010" as the "China National Committee for Biodiversity Conservation", and designated the Vice Premier of the State Council in charge of environment as the Director of the Committee. At present, this Committee has 23 member departments/ministries. It is mandated to coordinate all biodiversity conservation efforts in China and direct the implementation of China actions for UN Decade on Biodiversity. The establishment of the China National Committee for Biodiversity Conservation shows China's determination to strengthen environmental protection and promote sustainable development and its commitment to the international community. Since 2015, China has promulgated or revised 56 policies, laws and regulations related to biodiversity conservation, which gradually improve the policy and legislative system for biodiversity conservation in China.

In addition to the governance structure at central level, the environmental protection agencies of the governments at provincial level have also been carrying out reforms to better protect biodiversity. In 2008, SEPA was upgraded to the Ministry of Environmental Protection (MEP) and became a department of the State Council. All the provinces, autonomous regions and municipalities have upgraded their environmental protection agencies to departments, building a unified environmental protection system. Referring to the responsibility and function orientation related to biodiversity conservation at the national level, some provincial governments have established relevant coordination mechanisms, specifying the leading role of environmental protection departments in biodiversity conservation and the corresponding responsibilities of multiple internal agencies within those departments. To go with the actual local conditions, some provinces have set up administration agencies in line with the needs of local biodiversity conservation. For example,

Yunnan Province has set up a Lake Protection and Administration Division, showing the local features of its institutional reform and biodiversity conservation. In 2018, in accordance with the *"Decision of the Central Committee of the Communist Party of China on Deepening the Reform of the Party and Government Institutions"*, the Chinese State Council established the new MEE to practice the holistic thinking of integrated management of mountains, waters, forests, farmlands, lakes, grasslands and sand. All provinces, autonomous regions and municipalities have formed their new Department of Ecological Environment to comprehensively guide, coordinate and supervise the work of eco-environmental protection.

Annex 4 gives a more detailed description of China's progress in implementing policies for biodiversity conservation, elements of corresponding governance structure, as well as additional major institutions with conservation and management responsibilities which includes the China National Committee for Biodiversity Conservation and the Biodiversity Committee of the China Academy of Sciences.

4.1.2 Green Belt and Road and Biodiversity Conservation

In the light of China's experience in biodiversity conservation and need for building a green BRI, early efforts to align green BRI and biodiversity conservation have seen growth in the areas of governance mechanisms, governance system, information, technology development and scientific research, as well as green investment and finance, so as to jointly promote biodiversity conservation and implementation of SDG 15 in BRI participating countries.

First, there was a focus on establishing a mechanism and platform for cooperative governance to enable the improvement of a governance system for biodiversity conservation in BRI participating countries. Important progress began to show in integrating the existing bilateral and multilateral international cooperation mechanisms with green Belt and Road, building a network for biodiversity conservation, innovating cooperative models, as well as formulating a cooperation platform with inclusive participation of multiple stakeholders, including national governments, think tanks, business, civil societies and the wide participation of the public. Meanwhile, it is necessary to give full play to the mechanisms established for China-ASEAN cooperation, the Shanghai Cooperation Organization (SCO), the Lancang-Mekong cooperation, the Conference on Interaction and Confidence Building Measures in Asia, Euro-Asia Economic Forum, the Forum on China-Africa Cooperation, and the China-Arab States Cooperation Forum among other cooperative platforms. Efforts are still needed to facilitate the establishment of environmental cooperation platforms for the six major Economic Corridors and expand cooperation with relevant international organizations and agencies, so as to promote the effective implementation of SDG 15.

Second, efforts have been made to enhance cooperation on green technologies and research and development. There is a growing demand for the transfer of green, advanced and applicable technologies in developing countries along the Belt and Road, as well as need for joint research and development, promotion and application of cutting-edge technologies on the conservation of biodiversity. Specifically, further actions are to be taken for a platform on scientific research and technology development across scientific and research institutions and think tanks. Joint research with relevant countries and regions on biodiversity is a favorable opportunity for the conservation of global biodiversity. With the scientific study over the biodiversity of countries and regions along the Belt and Road, it would contribute to the analysis on the biodiversity evolution mechanisms and its characteristics and patterns on geographical distribution in these regions, expedite the scientific research on global diversity and help to provide training and capacity building for young officials and scientists in countries along the Belt and Road.

Third, steps are emerging to promote information exchanges, including biodiversity-related information sharing and disclosure, as well as provision of comprehensive information as decision-making support and safeguard. Growing needs are observed for enhancing the construction of biodiversity information base on the BRI Environmental Big Data Platform; for the full inclusion of national spatial and information infrastructure; for the exchange and sharing on environmental laws and regulations, policy standards and practices and experience; for enhanced comprehensive cooperation among different national departments and the sharing and disclosure on the ecological and environmental information; and for the improved capacity on risk evaluation and prevention targeting at BRI projects overseas. It is necessary to facilitate cooperation on the ecological and environmental information products, technologies and service to provide comprehensive information support and safeguard for building a green BRI.

Fourth, promising progress has been observed in the development of systems on green investment, green trade and green finance. Green finance systems help to build up the foundation for the long-term run of BRI projects. A good example is the *Social Responsibility and Environmental Protection Guidelines for Investments in the ASEAN Region* released by China-ASEAN Investment Cooperation Fund (CAF). The document prescribes that when CAF provides consulting services for businesses on overseas investment based on its Environmental and Social Management System Arrangements (ESMS), it could refer to the Performance Standards to identify and manage the impacts of environmental and social risks, clarify the evaluation metrics during the investment process and continuously monitor the later-stage investment, as a way to facilitate invested enterprises to avoid, ease and manage risks via a sustainable operation way. This Performance Standards covers eight areas including biodiversity conservation and the sustainable management of biological and natural resources, which jointly composed the standards that clients should meet on sustainable management of biodiversity through the overseas investment process. Specifically, it includes: ① checking whether the company understands and deals with the impacts of the project on biodiversity; ② checking whether the company carries out activities in regions under legal protection; ③ checking whether alien species are introduced in the process of project execution, and checking whether the company has the approval or permission from competent authorities if there are plans on introducing alien species; ④ checking whether the natural resources, forest and vegetation, fresh water and marine resources utilized by the project can be regenerated and whether the company is dedicated to managing them in a sustainable way.

Fifth, such documents as *the Green Investment Principle* (GIP) for BRI have been released to enhance green guidance on business activities and encourage businesses to adopt voluntary measures for environmental protection and sustainable development; it is important to motivate environmental business to explore the national markets in BRI participating countries and guide competitive environmental companies to "go global" in clusters with reference to China's experience and standard in building demonstrative ecological industrial parks, so as to enhance biodiversity conservation, prioritize the in-situ conservation and protection in proximity and take actions for ecological restoration. Meanwhile, efforts have been taken to guide business to augment the research, development and application of major technologies in addressing climate change.

Sixth, there is an increasing necessity to promote gender equality in the BRI cooperation and strengthen female leadership in biodiversity conservation. Biodiversity and gender are hot topics at the international level. Promoting gender mainstreaming in biodiversity conservation has gained widespread attention from the international community in recent years. Biodiversity and gender have been included in the CBD as a key issue. However, such problems as imperfect mechanisms and weak awareness related to gender exist in biodiversity research in China. In view of such problems, the following steps are thus recommended: set up gender focal points in all departments and establish a cross-sectoral communication and cooperation mechanism for gender mainstreaming to comprehensively enhance institutional capacity building; conduct gender mainstreaming training in biodiversity management departments and institutions to raise basic awareness of staff; as well as consider gender in the policies related to eco-environmental protection and green Belt and Road development and set up gender indicator in the evaluation system for specific projects. Such practice will also help BRI projects to meet the gender-related international standards and requirements of the host countries, promote people-to-people bond and enable the development of BRI to move forward steadily.

4.2 Survey and Assessment of International Practices

As mentioned in Section 2.2, the environmental management in the international systems has evolved rapidly over the last few decades. The same can be said for the enforcement and accountability of those systems. Just as Section 2.2 profiled the screening and assessment systems of international DFIs, this section profiles the accountability mechanisms of those same bodies.

Across the world, DFIs have mobilized to address SDG 15 and ensure that their activities protect project-affected biodiversity. While section 2.2 explained standards and guidelines, this section explains the DFI governance structures that have been adapted to ensure that conservation is sufficiently considered. It compares governance structures as adopted by Chinese policy banks' peers: major DFIs, both multilateral as well as national in nature. It includes: AFDB, ADB, AIIB, EBRD, EIB, IDB, IFC, KFW, and WB.

4.2.1 Governance for biodiversity: Incorporating SDG15 into DFI decision-making

As explained in Section 2.2, most major DFIs incorporate biodiversity considerations into their operations through the use of set standards, mitigation hierarchy deployment and consultations with affected stakeholders, who are likely to depend upon the local ecosystem for their livelihoods and therefore be particularly attuned to any biodiversity threats. In addition to these processes, several DFIs also incorporate other steps to mainstream SDG 15. These approaches are varied across DFIs. However, commonalities do arise in the requirements that DFIs set for themselves in this aspect, including:

- Incorporating expertise into assessments: the AFDB and AIIB require input from qualified experts to identify potentially-impacted ecosystems and ecosystem services.

● Empowering project implementers to adapt to changing conditions: AFDB, AIIB, EIB and the World Bank all require the use of adaptive management in their projects. In this approach, borrowers and clients must allow for the possibility that as they develop their projects, conditions will not be what they initially expected. Newly-discovered species or other biodiversity-related project impacts may emerge. Project plans should specify what types of challenges may arise and how the project implementers will adapt to these changing circumstances. With this planning done, implementers are empowered to change plans during the course of the project. In the case of the AIIB, major changes require additional environmental assessments to ensure that they are adapting their plans adequately.

4.2.2 Policy implementation: Monitoring and reporting

Borrowers and clients may commit to responsible environmental management, and DFIs strive to consider the implications for biodiversity, but actual performance will determine final outcomes. To this end, DFIs often institute monitoring and reporting requirements for their borrowers and clients. In doing so, DFIs often emphasize their respect for the national sovereignty of borrowing nations, devising methods that prioritize collaboration between lender and borrower for the best possible outcomes. Several different approaches emerge among DFIs, ranging from those that give borrowers the most responsibility in monitoring to those that utilize outside auditors.

4.2.3 Policy implementation: Grievance mechanisms

Many DFIs—multilateral as well as national—have instituted policies for stakeholders, including independent NGOs and project beneficiaries, to file grievances and request an investigation if they suspect that biodiversity has been harmed in the pursuit of DFI-supported projects. By developing institutional mechanisms for hearing, investigating and ruling on these claims, DFIs can ensure that their borrowers and grantees are living up to the terms of the agreement, prevent small harms from ballooning into larger harms, protect their own reputation globally, learn from their experiences and incorporate these lessons into future activities.

These grievance mechanisms can be at the DFI level, the project level or both. *Project-level grievance mechanisms* allow greater flexibility, by promoting the resolution of problems in a way that is often faster and more accessible for stakeholders than relying on one centralized system for claims from projects all around the world. However, they can be more cumbersome for DFIs to manage, requiring oversight of processes in many different countries. Table A4-1, in Annex 4, describes common elements in the design of project-level grievance mechanisms: their design, institutional location, processes and treatment of claimants.

DFI-level grievance mechanisms allow for stakeholders to bring a claim to the central DFI body or its designated complaint mechanism for consideration. These mechanisms can be simpler to manage for DFIs, as they only entail the creation and management of one body. However, they can be less accessible for project-affected stakeholders, and may mean that some rulings take more time than they would in project-level mechanisms.

Table A4-1 (in Annex 4), shows the various policy elements that DFIs incorporate into their project-level grievance mechanisms. A wide variety of arrangements exist, enabling DFIs to learn from these examples in designing their own mechanisms.

All of the DFIs listed in Table A4-1 (in Annex 4) also have DFI-level grievance mechanisms, though their design is too varied to display in table form. In addition to these DFIs, several other major multilateral and national development banks have these mechanisms, including the IDB, CAF, and BNDES.

4.2.4 Incorporating Gender

Regardless of the venue used, international DFIs have learned the importance of ensuring that accountability mechanisms are accessible for women. In many rural, poor settings, women's property rights are limited, such that ownership is recorded through their fathers, husbands or sons. In these contexts, national justice systems may not recognize their standing to bring a complaint through local courts, as they may not be able to demonstrate a loss to the value of their property. However, if their concerns are not heard, gender-based biodiversity risks may be unheeded and worsen. Both the ADB and the World Bank have recommended that their projects ensure accessibility for women to their accountability mechanisms, regardless of property[17,62]. This stage completes the upstream-to-downstream inclusion of gender considerations in biodiversity finance, so that it conld ensure that women are not disproportionately impacted in ways that can limit their ability to act as biodiversity stewards at the local level. Table 4-2 collects best practices from international DFIs on incorporating gender throughout the entire project cycle. It is not intended to be a comprehensive list but rather a collection of common best practices as recorded by research and evaluation staff at DFIs worldwide.

Table 4-2 Best Practices in Incorporating Gender into Biodiversity Finance

Project stage	Best practices
Upstream: planning	In planning for expected local biodiversity losses and changes to community access to local ecosystems, disaggregate the expected impact on local livelihoods by gender. Ensure that women are not disproportionately hurt by greater difficulty in carrying out traditional gathering roles. This practice is particularly effective in contexts where women and men have different traditional work roles. In arranging stakeholder engagement processes, ensure that women can participate fully. This practice helps planners understand the potentially different ways that a project may impact men and women differently. In contexts where women do not traditionally participate in mixed-gender public discussions, consider designing women-only engagement spaces.
Midstream: implementation	In projects where communities receive monetary compensation for a loss of access to local ecosystems, ensure that the financial compensation is distributed in such a way that it does not worsen women's well being. This practice is particularly relevant in contexts where women traditionally control resources they gather from local ecosystems but men control financial resources.

Project stage	Best practices
Downstream: monitoring and accountability	Account for changes in men's and women's use of time as well as financial resources. In contexts where women serve as local stewards of crop biodiversity through the cultivation of heirloom crop varieties in household or village gardens, this practice can ensure that biodiversity does not suffer. Garden crop biodiversity can be the key to the resilience of local food systems during extreme weather events or economic turmoil. Ensure that accountability and grievance mechanisms are fully accessible to women. This practice is particularly important in contexts where women lack equal property rights, have limited access to local judicial systems, or do not traditionally participate in mixed-gender public discussions. Women's participation in accountability mechanisms can allow project overseers and sponsors to monitor impacts on women's traditional role of crop biodiversity caretaker. As part of the post-project evaluation, develop a "tip sheet" for incorporating gender into future project planning in this particular context. This running collection of wisdom will help ensure that future development projects in this cultural context will be able to fully incorporate lessons learned through this project.

4.3 Areas of Convergence and Divergence between China and International Peers

Project governance structures entail both administrative and enforcement mechanisms, both within China and internationally. Domestically, administrative measures entail collaboration among ministries and other government bodies, while enforcement is the purview of the judicial system. Internationally, China's peers collaborate in similar ways by harmonizing standards and expectations among actors internationally. International enforcement and grievance mechanisms serve similar functions to domestic courts, in a way that enables participation from all parties and prioritizes dispute resolution.

Administratively, China's biodiversity-related governance is led by the MEE. In addition, the State Council has formed the National Committee for Biodiversity Conservation to oversee biodiversity actions at the national level, including representation from 23 ministries and institutions including MEE. Such institutional arrangements have incorporated biodiversity into socio-economic development and sectoral management routines, substantially boosting biodiversity mainstreaming in China. Beyond the national level, provinces, municipalities and autonomous regions have upgraded their environmental protection agencies to departments, building a nationally harmonized network.

The Chinese judicial system is the primary venue for policy implementation and enforcement, as is to be expected in a domestic context. The Supreme People's Court of China has established an environmental resources division, with guidelines for special biodiversity-related investigations and trials, for courts at all levels across ecosystems. These guidelines serve the goal of improving environmental dispute resolution throughout China, in a unified manner.

Internationally, multilateral government bodies and development finance institutions have incorporated biodiversity governance into project management in a broad spectrum of ways: incorporating feedback from project stakeholders and independent experts; empowering project managers to adapt to changing conditions; establishing institutional mechanisms for accountability and handling grievances; incorporating gender as a crucial aspect of consideration throughout the project cycle, among others. Administrative mechanisms take the form of upstream planning, incorporating compulsory standards, the mitigation hierarchy and stakeholder participation measures. As China's international peers include both multilateral and national institutions, and governments as well as financial institutions, harmonized environmental and social risk management approaches have emerged to ensure well-calibrated expectations by all parties from the point of project initiation, to protect collaborative relationships as well as ecosystems throughout the project cycle.

As the BRI continues to expand, harmonized expectations will continue to grow in importance as well. The collaborative nature of the BRI allows for relevant Chinese ministries to take active roles in setting standards, in partnership with peer government representatives from along the Belt and Road.

DFIs that are peers to China's main lenders for BRI projects (the China Development Bank and Export-Import Bank of China) have developed a wide array of monitoring and grievance mechanisms to ensure that their administrative measures are effective. These measures are collaborative by definition and serve as dispute resolution fora for local communities, national host country governments and implementing contractors. Like domestic courts, they reinforce public trust in the governance institutions involved. The BRI could benefit strongly from the design and incorporation of a similar environmental dispute resolution mechanism, with a particular focus on biodiversity concerns.

5 POLICY RECOMMENDATION: CONSTRUCTION ROADMAP OF GREEN BELT AND ROAD

In the previous chapters the research team has described the progress that has been made within China and among China's peers in balancing the benefits of investment with the risks to communities and the ecosystems that support them. Given the speed and scope of this institutional progress, it is crucial to harness all of the lessons possible for additional growth, in order to ensure that the BRI fulfills its potential to support sustainable development globally. Built on the results of the first phase of SPS on Green Belt and Road and 2030 Agenda for Sustainable Development, this chapter further improves the roadmap for building a green BRI and proposes policy recommendations for aligning BRI and SDG 15.

5.1 Roadmap for Building a Green BRI

5.1.1 Enhancing policy communication, taking the green Belt and Road Initiative as an important practice of realizing Sustainable Development Goals and facilitating global environment governance reform in 2030

It is important to set green development as the fundamental principles of building the Belt and Road. China has the opportunity to integrate green development and ecological civilization through the "five connectivities" in building the Belt and Road, facilitate the construction of green infrastructure, green investment and green finance, and build the Belt and Road into one route towards green and sustainable development so as to establish a community with shared future for mankind on the basis of green development.

China should augment cooperation in the domain of environmental protection on international multilateral platforms for BRI. Specifically, it is necessary to incorporate the Thematic Forum on "Green Silk Road" as a fixed thematic forum within the schedule of the Belt and Road Forum for International Cooperation (BRF). It is necessary to bring into full play the role of the BRI International Green Development Coalition and the Belt and Road Sustainable Cities Alliance in serving as the international platforms for jointly developing the Green Silk Road, facilitating the realization of SDGs and improving global environmental governance system. Efforts are recommended to disseminate the concepts and practice of green development in BRI participating countries via champion countries, pilot cities and demo projects. In addition, China should make good use of BRI's strengths in the "five connectivities" to jointly facilitate the implementation of policies on ecological and environmental protection, biodiversity conservation, climate change mitigation and adaptation etc. to bolster support for existing international conventions such as CBD, *the Convention on International Trade in Endangered Species of Wild Fauna* and *Flora* (CITES), UNFCCC, etc.

5.1.2 Enhancing strategic alignment, establishing the mechanism for linking Green BRI with the 2030 Agenda for Sustainable Development

Given that a green BRI is a crucial tool in realizing the 2030 Agenda for Sustainable Development and in particular promoting international biodiversity conservation, this SPS recommends the following steps to strategically align planning with biodiversity goals.

Enhance policy design. This report recommends that China incorporates the implementation of the 2030 Agenda for Sustainable Development (SDGs) as an important task into building a green Belt and Road. When signing MOUs on jointly building the Belt and Road with relevant countries and international organizations, China needs to include jointly building a green Belt and Road and expediting the alignment of BRI and the 2030 Agenda as an important part of these MOUs.

Establish a mechanism for implementation. This report recommends setting up working groups/expert teams with partners based on the situations in different countries and jointly draft strategies for building a Green Silk Road together, identifying the priority areas for cooperation in both the short and long terms and fostering sound linkages between different national plans based on the practical needs of implementing SDGs in BRI participating countries.

Develop mechanisms for participation and feedback. A network should be built with government guidance, business support and public participation, prioritizing the perfection of mechanisms on the involvement of international organizations. Initiate the mechanism for whole-process participation, covering negotiation, decision making and dynamic feedback, in order to make sure the successful alignment of building green Belt and Road and implementing the 2030 Agenda for Sustainable Development under open and transparent circumstances.

Establish professional mechanisms for cooperation for cities and localities along the Belt and Road. Cities along the Belt and Road should be encouraged to consider their own industrial structure, advantages and development goals, and create a policy framework favorable for addressing issues of common concern to explore opportunities for cooperation and guide private sector in participating BRI cooperation.

5.1.3 Improving project management, and establishing and improving the mechanism for project management on green Belt and Road

To incorporate the above-mentioned strategies into BRI project management, this SPS recommends the following steps.

Establish a mechanism for eco-environmental management of BRI projects. It is important to strengthen communication and coordination between China and BRI participating countries and that among different Chinese government agencies. China should establish science-based risk evaluation and management mechanisms for their projects to respond to various risks, strictly follow the host countries' norms and standards in such procedures as project design, construction, operation, procurement and bidding. An encouraging environment should be created for BRI projects to apply the principles, standards and customary practices for environmental protection that are used by international organizations and multilateral financial institutions, and strive to realize goals that are made with high standards, beneficial for people's livelihoods and sustainable. China needs to support its financial institutions to incorporate the ecological and environmental impacts of projects as a key factor in their project rating and risk rating systems, and put forward evaluation methods and instruments on the environmental and social risks for BRI projects as an important metric for granting govermental support, development financial support and policy financial support. Practice in commercial finance is encouraged to adopt similar standards.

Call for wide application of green finance instruments under the Belt and Road framework. First, establishment of the Belt and Road Green Development Fund needs to be explored, with priority given to projects in support of the development of ecological and environmental infrastructure, capacity building and green industries in countries along the Belt and Road; Second, it is necessary to establish guarantee agencies on green investment and financing under the BRI with the wide participation of different countries, in order to share risks and mobilize social capital in green domains; Third, there is a need to establish the mechanism for environmental information disclosure and enhance the transparency of information based on the BRI Environmental Big Data Platform.

Speed up facilitation of trade in environmental products and services. Improve the opening level of environmental products and services market, encourage enhanced import and export of environmental products and services such as pollution prevention and treatment technologies and services, and help foster green industrial development in BRI participating countries.

5.1.4 Improving capacity building, and jointly conducting green capacity building programs with BRI participating countries

With regard to public engagement, this SPS recommends that BRI project planners take the following steps.

Enhance people-to-people bonds among BRI participating countries. The Green Silk Road Envoys Program should be expanded into a flagship program on capacity building under the Belt and Road framework, which aims to enhance ecological and environmental cooperation and communications and share the ideas and practices of building an Ecological Civilization and green development in China via such activities as capacity building workshops for environmental officials, managers, and practitioners, consultation for policy development, etc.

Support and facilitate the exchange and cooperation of environmental organizations from China and BRI participating countries. We should clarify the leading and responsible government department, and then guide/support environmental organizations to build up their own cooperation networks; increase the government efforts to purchase environmental protection social organization's services, set up special cooperation funds to support environmental protection organization to go out; In addition, efforts are needed to perfect the mechanism for the involvement of environmental social organizations and come up with a list of items on international communication with the participation of environmental organizations.

Facilitate gender mainstreaming and augment women's leadership roles. There is a necessity to improve gender consciousness among policy makers and women communities and facilitate the mainstreaming of gender consciousness in the process of policy formulation and project implementation for building a green Belt and Road; This report recommends enhancing institutional capacity building on gender mainstreaming in environmental protection related agencies and exploring the possibility of establishing a cross-sectoral communication mechanism to facilitate gender mainstreaming; With the help of the Green Silk Road Envoys Program, China could organize thematic capacity building programs and seminars on improving women's leadership roles in green development upon inviting the participation of female officials, experts, scholars and youth in the domain of environmental protection from BRI participating countries, and share methods and experience in gender mainstreaming with BRI partners.

5.2 Policy Instruments for Aligning the BRI with SDG 15

Under the framework and in the spirit of the general roadmap for building a green BRI outlined above, this session recommends the policy directions for aligning BRI, SDG 15 and CBD. The proposed policy recommendations are built with full consideration of the main objective and approaches internationally used for biodiversity conservation. The main objective is focusing on the establishment and mainstreaming of global standards, which are primarily practiced through the establishment of operational risk management strategies to protect institutional reputations and the cooperative relationships. The corresponding major approaches for operationalizing standards include: ① aligning institutional practices with international or national commitments; ② using exclusionary lists of categorically ineligible projects; ③ requiring projects to undergo biodiversity impact assessments; ④ adopting a mitigation hierarchy to do no harm and if possible benefit local ecosystems; ⑤ incorporate local stakeholders. The following five policy directions are thus proposed.

First, apply international norms and standards to facilitate the use of stricter environmental standards in BRI projects. It is recommended to actively align BRI efforts with the fulfillment of international and national commitments to international conventions, including *the Convention on Biological Convention* and *the United Nations Framework Convention* on Climate Change. It is also necessary to align BRI with other biodiversity related international conventions that China is a signatory to such as *the International Convention for the Protection of New Varieties of Plants*, *Convention Concerning the Protection of the World Cultural and Natural Heritage*, *Convention on International Trade in Endangered Species of Wild Fauna and Flora*, *Convention on Wetlands of International Importance Especially as Waterfowl Habitat*, and achieve synergies with climate related conventions like UNFCCC.

The BRI projects should comply with the environmental laws, regulations and standards of the host country. These projects are encouraged to adopt environmental protection principles, standards and practices implemented by international organizations and multi-lateral financial institutions. In keeping with *the Green Credit Guidelines* issued by the China Banking Regulatory Commission (CBRC), environmental impact assessments should ensure that a proposed BRI project adheres to the host country's biodiversity standards, international conventions that China and the host country are parties to, and the principles for harmonizing the BRI with sustainable development outlined by the China Development Bank (UNDP-CDB, 2019). In the meantime, it is recommended to jointly promote green value chain under the framework of BRI, so as to accelerate the dual circulation pattern of domestic economic cycle and international economic cycle and push forward green recovery.

Second, focus on environmental impacts and carry out assessment and classification-oriented management of BRI projects. It is recommended to boost the development of the guidance on assessment and classification of BRI projects, which should include clearly defined positive and negative lists, in order to give adequate attention to the projects' potential impacts related to environmental pollution, biodiversity conservation and climate change based on the ongoing joint research on *Green Development Guidance for BRI Projects* undertaken by BRIGC, which could provide green solutions to BRI participating countries and projects and provide green credit guidelines to financial institutions. It is recommended to pilot concept and practice of green development via champion countries, pilot cities and demo projects and enhance BRI green development case studies and experience sharing. The research on *Green Development Guidance for BRI Projects* shows that the assessment and classification-oriented management should consider the various international and national commitments of the host countries and meet the host countries' needs for economic growth and environmental protection. Such management should guide and assist businesses to incorporate environmental impact assessment (EIA), as well as biodiversity conservation and impact mitigation measures at the stage of project design.

Third, improve policy instruments to enhance eco-environmental management of BRI projects. These instruments include the following four aspects.

(1) It is recommended to carry out environmental management assessment for key BRI sectors and projects and establish a regular environmental management regulatory mechanism that incorporates environmental pollution, biodiversity conservation and climate change as important factors for assessment. The biodiversity analysis should fully examine the ecological and socio-economic conditions of the project locality, gauge the direct, indirect and cumulative impacts of the project on wildlife habitats and biodiversity, and consider how the project-affected stakeholders value biodiversity and wildlife habitats.

(2) It is recommended to make full use of green finance instruments and environmental risk assessment tools, establish biodiversity conservation governance and financing framework and give full play to the proactive role of financial institutions in guiding green investment. As a great number of biodiversity hotspots are located in the developing world, it is difficult to depend on the design and execution of long-term projects that can only generate economic returns after many years to provide necessary conversation funds. Thus there is emerging need to carry out bilateral and multilateral cooperation on biodiversity financing. Over the past decade, biodiversity financing in China and the rest of the world has been booming substantially; while rich experience has been accumulated in improving ecological compensation schemes, transfeing payments to ecological function areas and subsiding for returning farmland to forests program. Given that biodiversity conservation requires an enabling policy environment, the MEE should be charged to work with the NDRC and other administrative bodies to design biodiversity impacts mitigation strategies and to jointly design financing mechanisms for mitigation, compensation, and restoration schemes in consultation with various stakeholders in the Chinese government, host countries and other affected parties and partners. Besides, it is recommended that China increase the share of environment aid in the official development aid to the BRI participating countries, so as to strengthen

biodiversity financing.

(3) It is recommended to take ecological redlining as a key instrument to align the BRI and SDG 15 and share best practices in ecological redlining with participating countries. It is important to support the BRI participating countries to build on the experience of ecological redlining in developing similar land use strategic planning.

(4) It is of great significance to establish a proper stakeholder consultation mechanism for BRI projects, with the purpose to guarantee the effective stakeholder participation and consultation in terms of biodiversity assessment and management.

Fourth, improve the coordination mechanism and facilitate effective linkage and alignment among different SDGs using Nation-based Solutions (NBS). It is important to create synergies with efforts for SDG 13 of Climate Action, consider a step-by-step reduction of investments in carbon-intensive industries such as coal-fired power plants to prevent carbon lock-in, and further strengthen investments in green projects on environmental protection and renewable energy to encourage environmentally-sustainable, green and low-carbon projects. The concept of green development should be incorporated into the selection, implementation and management of infrastructure projects with strengthened efforts being made in researching and developing the guidelines on sustainable infrastructure operation.

ANNEXS

Annex 1: Supporting Evidence for Chapter 1

Table A1-1 Geographic Distribution of Countries that have Signed BRI Memorandums of Understanding

Region	BRI Countries
East Asia	1. Mongolia, 2. Republic of Korea
ASEAN countries (10 countries)	1. Singapore, 2. Malaysia, 3. Indonesia, 4. Myanmar, 5. Thailand, 6. Laos, 7. Cambodia, 8. Viet Nam, 9. Brunei, 10. Philippines
West Asia (17 countries)	1. Iran, 2. Iraq, 3. Turkey, 4. Syria, 5. Jordan, 6. Lebanon, 7. Israel, 8. Saudi Arabia, 9. Yemen, 10. Oman, 11. UAE, 12. Qatar, 13. Kuwait, 14. Bahrain, 15. Greece, 16. Cyprus, 17. Sinai Peninsula of Egypt
South Asia (8 countries)	1. India, 2. Pakistan, 3. Bangladesh, 4. Afghanistan, 5. Sri Lanka, 6. Maldives, 7. Nepal, 8. Bhutan
Central Asia (5 countries)	1. Kazakhstan, 2. Uzbekistan, 3. Turkmenistan, 4. Tajikistan, 5. Kyrgyzstan
Commonwealth of Independent States (7 countries)	1. Russia, 2. Ukraine, 3. Belarus, 4. Georgia, 5. Azerbaijan, 6. Armenia, 7. Moldova
Central and Eastern Europe (16 countries)	1. Poland, 2. Lithuania, 3. Estonia, 4. Latvia, 5. Czech Republic, 6. Slovakia, 7. Hungary, 8. Slovenia, 9. Croatia, 10. Bosnia and Herzegovina, 11. Montenegro, 12. Serbia, 13. Albania, 14. Romania, 15. Bulgaria, 16. Macedonia
Western Europe (7 countries)	1. Austria, 2. Finland, 3. France, 4. Italy, 5. Luxembourg, 6. Malta, 7. Portugal
North Africa (5 countries)	1. Algeria, 2. Libya, 3. Mauritania, 4. Morocco, 5. Tunisia
West Africa (11 countries)	1. Cabo Verde, 2. Cote d'Ivoire, 3. Gambia, 4. Ghana, 5. Guinea, 6. Liberia, 7. Mali, 8. Nigeria, 9. Senegal, 10. Sierra Leone, 11. Togo
Central and Southern Africa(8 countries)	1. Angola, 2. Cameroon, 3. Chad, 4. Republic of Congo, 5. Equatorial Guinea, 6. Gabon, 7. Namibia, 8. South Africa
East Africa (15 countries)	1. Burundi, 2. Djibouti, 3. Ethiopia, 4. Kenya, 5. Madagascar, 6. Mozambique, 7. Rwanda, 8. Seychelles, 9. Somalia, 10. South Sudan, 11. Sudan, 12. Tanzania, 13. Uganda, 14. Zambia, 15. Zimbabwe

Region	BRI Countries
Latin America (11 countries)	1. Bolivia, 2. Chile, 3. Costa Rica, 4. Ecuador, 5. El Salvador, 6. Guyana, 7. Panama, 8. Peru, 9. Suriname, 10. Uruguay, 11. Venezuela
Caribbean (8 countries)	1. Antigua and Barbuda, 2. Barbados, 3. Cuba, 4. Dominica, 5. Dominican Republic, 6. Grenada, 7. Jamaica, 8. Trinidad and Tobago
Oceania (9 countries)	1. Fiji, 2. Kiribati, 3. Micronesia, 4. New Zealand, 5. Papua New Guinea, 6. Samoa, 7. Solomon Islands, 8. Tonga, 9. Vanuatu

Note: Timor-Leste is currently in the process of ASEAN accession.

Table A1-2 The state of BRI participating countries in implementing SDG 15

Region	Country	Goal 15 Implementation	Goal 15 Trend	Region	Country	Goal 15 Implementation	Goal 15 Trend
East Asia	Mongolia		↗	Central Asia	Turkmenistan		→
	Republic of Korea		→		Kazakhstan		→
ASEAN	Singapore		.		Uzbekistan		→
	Indonesia		↓		Kyrgyzstan		→
	Malaysia		→		Tajikistan		→
	Cambodia		↓	West Asia	Iraq		→
	Vietnam		↗		Kuwait		.
	Myanmar		↓		United Arab Emir.		.
	Lao P. D. R.		↓		Saudi Arabia		→
	Philippines		↓		Syria		→
	Thailand		→		Israel		↓
	Brunei		.		Yemen		↓
South Asia	Maldives		.		Oman		.
	India		↓		Turkey		→
	Afghanistan		↓		Bahrain		
	Bhutan		→		Lebanon		→
	Bangladesh		↓		Qatar		.
	Sri Lanka		↗		Iran		↓
	Pakistan		↓		Sinal Peninsula, Egypt		→
	Nepal		→		Cyprus		.

Green BRI and 2030 Agenda for Sustainable Development
—Aligning with Sustainable Development Goal 15 to Promote Global Biodiversity Conservation

Region	Country	Goal 15 Implemen-tation	Goal 15 Trend	Region	Country	Goal 15 Implemen-tation	Goal 15 Trend
West Asia	Greece		↗	Western Europe	France		↗
	Jordan		.		Finland		↑
Central and Eastern Europe	Montenegro		↓		Italy		↑
	Serbia		↗	North Africa	Algeria		→
	Bosnia and Herz.		→		Tunisia		↗
	N. Macedonia		↗		Mauritania		.
	Croatia		↗		Morocco		→
	Albania		↗		Libya		.
	Slovenia		↑	East Africa	Djibouti		↓
	Romania		↑		Madagascar		→
	Slovakia		↑		Ethiopia		→
	Hungary		↑		Tanzania		→
	Lithuania		↑		Kenya		↓
	Estonia		↑		Sudan		↗
	Czech Republic		↑		Rwanda		→
	Poland		↑		Mozambique		→
	Latvia		↑		Uganda		↗
	Bulgaria		↑		Zambia		→
Comm. of Indep. States	Georgia		→		Burundi		↑
	Armenia		↓		Zimbabwe		↗
	Ukraine		→		Seychelles		.
	Moldova		→		Somalia		→
	Russian.		→		South Sudan		↗
	Azerbaijan		→	Central and Southern Africa	South Africa		↗
	Belarus		↑		Angola		→
Western Europe	Luxembourg		↗		Cameroon		→
	Malta		.		Chad		↑
	Austria		↗		Gabon		↑
	Portugal		↗		Namibia		↑
	Senegal		.		Congo, Republic		↑
	Gambia		.		Equatorial Guinea		.
	Ghana		.				

ANNEXS

Region	Country	Goal 15 Implemen-tation	Goal 15 Trend	Region	Country	Goal 15 Implemen-tation	Goal 15 Trend
West Africa	Liberia		→	Caribbean	Jamaica		.
	Cabo Verde		.		Trin. & Tobago		.
	Sierra Leone		↑		Cuba		.
	Mali		→		Dominican Rep.		↗
	Cote d'Ivoire		↗				
	Nigeria		↗		Antigua & Barb.		.
	Guinea		↗				
	Togo		.		Barbados		.
Latin America	Uruguay		↓		Dominica		.
	Panama		.		Grenada		.
	Chile		↓	Oceania	Fiji		↓
	Guyana		.		Vanuatu		.
	Ecuador		→		New Zealand		↓
	El Salvador		→		P.N.G.		.
	Costa Rica		.		Kiribati		.
	Peru		.		Micronesia		.
	Suriname		↗		Samoa		.
	Bolivia		↗		Solomon Isl.		.
	Venezuela		↗		Tonga		.

Colors			Trend Arrows	
Green	Goal Achievement		↑	On track or Maintaining Achievement
Yellow	Challenges Remain		↗	Moderately Increasing
Orange	Significant Challenges		→	Stagnating
Red	Major Challenges		↓	Decreasing
	Data not avaioble		.	Data not avaioble

Annex 2: Supporting Evidence for Chapter 2

Table A2-1 Assessment of China's progress in implementing SDG 15

SDG15	Main work undertaken by China to achieve SDGs	Indicators	Overall assessments and trends
15.1 By 2020, ensure the conservation, restoration and sustainable use of terrestrial and inland freshwater ecosystems and their services, in particular forests, wetlands, mountains and drylands, in line with obligations under international agreements	Safeguarding the ecological water level of important wetlands and estuaries, protecting and restoring wetland and river and lake ecosystems, establishing systems of wetland protection and degraded wetland protection and restoration and promoting the rational use of wetlands; promoting the development of the legal system of terrestrial nature reserves and improving the level of protection and utilization of natural resources such as forests; and conducting river and lake health assessments to protect aquatic ecosystems.	National-level protected areas for aquatic germplasm resources	●
		Number of wetland parks	●
		Percentage of surface water bodies with good quality meeting Classes I -III standards	●
15.2 By 2020, promote the implementation of sustainable management of all types of forests, halt deforestation, restore degraded forests and substantially increase afforestation and reforestation globally	Carrying out large-scale land greening, strengthening the implementation of key afforestation projects, improving the natural forest protection system, comprehensively stopping commercial forest logging and protecting and cultivating forest ecosystems; improving the policy of returning farmlands to forests and grasslands and exploring the establishment of mechanisms for government-sponsored social services to carry out afforestation and forest protection.	Total forest stock	●
		Area of natural forests	●

SDG15	Main work undertaken by China to achieve SDGs	Indicators	Overall assessments and trends
15.3 By 2030, combat desertification, restore degraded land and soil, including land affected by desertification, drought and floods, and strive to achieve a land degradation-neutral world	Participating in demonstration projects aiming at land degradation neutrality goal under the United Nations Convention to Combat Desertification; promoting the comprehensive control of desertification, rocky desertification and soil erosion, preventing land degradation, continuously expanding the scope of desertification land management and strengthening the ecological protection and construction of desert areas.	Forest stock in key ecological project areas	●
		Grassland vegetation cover rate in key ecological project areas	●
		Area of desertified land	●
15.4 By 2030, ensure the conservation of mountain ecosystems, including their biodiversity, in order to enhance their capacity to provide benefits that are essential for sustainable development	Comprehensively improving the stability of mountain ecosystems and ecological service functions and building an ecological security barrier; constructing national forest germplasm resource banks and establishing a system of standardized germplasm resource conservation; scientifically optimizing the forest park management system and promoting the sharing and utilization of forest diversity resources.	Number and area of forest parks	●
		Total timber standing stock	●
		Area of natural forests	●
		National investments in ecological conservation	●
15.5 Take urgent and significant action to reduce the degradation of natural habitats, halt the loss of biodiversity and, by 2020, protect and prevent the extinction of threatened species	Implementing major projects for biodiversity conservation; strengthening the construction and management of nature reserves, and increasing the protection of typical ecosystems, species, genes and landscape diversity; increasing the investment in ecosystem protection and restoration and carrying out large-scale survey of baselines for species resources in the country; establishing a national biodiversity observation network.	Red List Index	●
		Living Planet Index	●

SDG15	Main work undertaken by China to achieve SDGs	Indicators	Overall assessments and trends
15.6 Promote fair and equitable sharing of the benefits arising from the utilization of genetic resources and promote appropriate access to such resources, as internationally agreed	Gradually establishing and improving laws and regulations on the protection and benefit sharing of genetic resources and promoting the proper access to genetic resources and the fair and equitable sharing and utilization; increasing funding for the conservation of biological genetic resources and participating in international cooperation in access to and use of genetic resources.	Indicators related to access to genetic resources and benefit-sharing	...
15.7 Take urgent action to end poaching and trafficking of protected species of flora and fauna and address both demand and supply of illegal wildlife products	Seriously implementing the Wild Animal Protection Law and speeding up the improvement of the National List of Key Protected Wild Animals; optimizing the national wildlife protection network, strengthening the import and export management of wild animals and plants, and cracking down on illegal trade in wild animal and plant products such as ivory; restoring and expanding the habitats of endangered wildlife and promoting international cooperation in wildlife conservation.	/	/
15.8 By 2020, introduce measures to prevent the introduction and significantly reduce the impact of invasive alien species on land and water ecosystems and control or eradicate the priority species	Actively participating in international conventions related to the prevention and control of invasive alien species; improving the list of IAS and related risk assessments.	Number of newly discovered IAS every decade	●
		Batches and number of species of harmful pests intercepted at ports	●
		Number of IAS risk assessment standards released.	●
15.9 By 2020, integrate ecosystem and biodiversity values into national and local planning, development processes, poverty reduction strategies and accounts	Requiring governments of all levels to undertake ecological conservation and biodiversity conservation taking into account their local circumstances, and to incorporate biodiversity into their long-term and medium term development planning.	Number of sectoral policies related to conservation and sustainable use of biodiversity	●

SDG15	Main work undertaken by China to achieve SDGs	Indicators	Overall assessments and trends
15.a Mobilize and significantly increase financial resources from all sources to conserve and sustainably use biodiversity and ecosystems	Strengthening coordination and increasing funds needed for infrastructure and capacity building	National investments in ecological conservation	●
15.b Mobilize significant resources from all sources and at all levels to finance sustainable forest management and provide adequate incentives to developing countries to advance such management, including for conservation and reforestation	Promoting diversified resource mobilization strategies, guiding enterprises and the public to participate more deeply, and forming a long-term financial mechanism for forest management; helping other developing countries to carry out technical training under the framework of South-South cooperation to improve the rate of utilization of forest resources and the level of forest management; and guiding Chinese companies to carry out sustainable forest management and business operation abroad.	Ecological compensation for forest ecological benefits	●
15.c Enhance global support for efforts to combat poaching and trafficking of protected species, including by increasing the capacity of local communities to pursue sustainable livelihood opportunities	Strengthening the review of trade in species restricted by the international trade conventions in which China participates, and strictly managing the certification under the Convention on International Trade in Endangered Species of Wild Fauna and Flora; carrying out special actions to curb the criminal momentum of poaching and illegal trade of wild animals; encouraging and guiding the development of wild plant artificial cultivation industry.	Number of illegally smuggled or trafficked protected species intercepted or detected	…

● Status improving; ● Status worsening;
… no adequate data; / no indicators available for assessment.
Note: China's Sixth National Report on Implementation of the Convention on Biological Diversity. 2018.

Table A2-2 Operational Requirements for Biodiversity Safeguards Applied by DFIs to Clients

	ADB	AFDB	AIIB	BNDES	CAF	EBRD	EIB	IDB	IFC	KFW	WB
Screen and categorize projects for level of impact and risk to biodiversity	X	X	X	X	X	X	X	X	X	X	X
Assess baseline conditions	X	X	X		X	X	X	X	X		X
Assess direct, Indirect, cumulative and Induced impacts and risks to biological resources	X	X	X		X	X	X	X	X		X
Consider trans-boundary Impacts	X	X	X			X	X	X	X		X
Socio-economic impacts of modifications to biodiversity	X	X	X		X	X	X	X	X	X	X
Use of strategic environmental assessment	X	X	X				X	X	X		X
Apply the precautionary approach or principle	X		X		X	X	X	X		X	X
Examine alternatives to project design technology and components	X	X	X		X	X	X	X	X		X
Explicitly incorporate costs of environmental mitigation measures into environmental assessment					X			X			
Apply mitigation hierarchy		X	X	X		X	X	X	X		X
Explicit adherence to national law and host country international commitments	X	X	X	X	X	X	X	X	X		X
Option to use country and/or client systems in lieu of DFI safeguards	X		X					X			X
Engage independent experts and advisory panels		X	X			X					X
Carry out stakeholder consultation during environmental assessment and project implementation	X	X	X		X	X	X	X	X	X	X
Require client to disclose environmental assessments and management plans	X	X	X			X	X		X		X
Prepare Biodiversity Management or action plans			X			X	X				

	ADB	AFDB	AIIB	BNDES	CAF	EBRD	EIB	IDB	IFC	KFW	WB
Enhance Biodiversity	X	X	X		X	X	X		X		X
Use adaptive management procedures to address unanticipated impacts			X				X				
Criteria for Projects in/affecting Critical Habitat	X	X	X		X	X	X	X	X		X
Criteria for Projects in/affecting legally protected and internationally recognized areas	X	X	X	X	X	X	X		X		X
Criteria for Projects in/affecting natural habitat	X	X	X		X	X	X		X		X
Criteria for Projects in/affecting modified habitat	X				X		X		X		X
Use of Offsets	X		X			X	X		X		X
Management of Ecosystem Services		X			X	X	X		X		X
Sustainable management of natural Living and renewable resources	X	X			X	X	X		X		X
Control of Invasive Alien Species	X	X	X			X	X		X		X
Genetically Engineered Organisms		X					X				
Environmental Flows		X*									
Forest Management		X	X		X					X	
Marine Environment		X	X			X				X	
Protection of Indigenous Knowledge and commercial activities		X			X				X	X	X
Supply Chain Management		X				X	X		X		X
Impact of Climate Change on Biodiversity		X				X	X				
List of Categorically Ineligible Projects	X	X		X	X	X	X	X	X	X	X

X: Refers to projects that affect water resources.

Source: Web pages, official policies, and interviews with individuals at listed international institutions.

Annex 3: Evidence from Chapter 3

Detailed descriptions of conservation finance initiatives in China

Increasing transfer payments to ecological function areas. Since 2018 when the central government established the transfer payment system for key national ecological function areas, China has been intensifying the efforts to protect those areas. In 2018, the state made a transfer payment of 72.1 billion yuan to key national ecological function areas, which is 9.4 billion more than it did the previous year, registering an increase of 15%. Meanwhile, China has kept expanding the scope of key national ecological function areas. Once included in the scope, the area will receive financial and policy support as long as it strictly implements the negative list system for industrial access. According to relevant regulations, a region counted as a key national ecological function area needs to strengthen ecological protection and restoration, regulate the boundaries of industrialization and urbanization and enhance the supply capacity of eco-products.

Strengthening fiscal support from the central government to forestry ecological protection. On July 27, 2018, the Ministry of Finance and the State Administration of Forestry and Grassland jointly issued the *Management Measures for Forestry Ecological Protection and Recovery Funds* (*Measures*), aiming at regulating the management of Forestry Ecological Protection and Recovery Funds, coordinating the integrated use of such funds, improving the efficiency of utilization and facilitating forestry ecological protection and recovery. According to the *Measures*, Forestry Ecological Protection and Recovery Funds refer to special transfer payment funds in the central budget for the social insurance and social expenditure of Natural Forest Protection Project (hereinafter referred to as "NFPP"), the cessation of commercial clear-cutting of natural forest, improving relevant policies on returning farmland to forestry and initiating a new round of returning farmland to forestry and grassland. In 2018, a total of 41.604 billion yuan was allocated to several provinces, of which

Heilongjiang received the most, 8.595 billion. The *Measures* has clearly stated that the funds are allocated based on the factor method. The standard of cash subsidy for returning farmland to forestry is as follows: for the Yangtze River Basin and southern areas, 125 yuan per mu each year; for the Yellow River Basin and northern areas, 90 yuan per mu each year. Those returned eco-forests will be subsidized for 8 years, and those returned economic forests for 5 years. As for the new round of returning farmland to forestry and grassland, the returned forests will receive a cash subsidy of 1,200 yuan per mu, paid at 3 intervals within 5 years, with 500 yuan in the first year, 300 yuan in the second year and 400 in the third year; the returned grasslands will receive a cash subsidy of 850 yuan, paid at 2 intervals within 3 years, with 450 yuan in the first year and 400 yuan in the second.

Strengthen fiscal support from the central government to ecological protection and restoration of wetlands. From 2013 to 2016, the central government allocated 5 billion yuan to protect wetlands in China and continued to provide support through the Funds for Reform and Development of Forestry afterwards. The measures taken include: supporting the protection and restoration of wetlands, supporting returning farmland to wetland and supporting the wetland ecological benefit compensation.

Promote the grassland ecological protection subsidy incentive policy continuously. Since 2011 when the state implemented the grassland ecological protection subsidy incentive policy in 8 major pastoral areas in Inner Mongolia, Xinjiang, Tibet, Qinghai, Sichuan, Gansu, Ningxia and Yunnan and Xinjiang Production and Construction Corps, and gave out a total of 13.6 billion yuan as subsidies, 36 pastoral and agricultural pastoral regions in 5 non-major pastoral provinces including Heilongjiang have been added to the scope, altogether covering 268 pastoral and mixed farming-pastoral counties. In recent years, the state together with the General Bureau of Land Reclamation of Heilongjiang has implemented the grassland subsidy incentive in 13 provinces including Shanxi and Hebei and production and construction corps, achieving remarkable results in improving the grassland ecosystem, the production of animal husbandry and the life of herders. In

2018, a new round of grassland ecological protection subsidy incentive of 18.76 billion yuan was included in the central budget to support the banned grazing area of 1.206 billion mu and the grass-animal balance area of 2.605 billion mu, and award those regions with outstanding performance. The funds were utilized by local governments in grassland management and the transformation and upgrading of the production mode. Besides, the subsidies for banning grazing and incentive for grass-animal balance were required to be given out based on the principle of "to clear targets in a reasonable amount accurately", making sure each target could get their share in time. The distribution of the funds is publicized at the village-level, accepting surveillance by the masses. In addition to supporting the implementation of subsidies for banning grazing and incentives for grass-animal balance, the performance appraisal also requires no less than 70% of the funds should be used in protecting the grassland ecosystem and developing grass-based livestock husbandry, that relevant trails should be conducted in accordance with local realities and that support to new agricultural operators should be enhanced concerning the development of modern grass-based livestock husbandry.

Launching pilot programs on the unified confirmation and registration of natural resources. The confirmation and registration of natural resources is important to promoting the reform of the property right mechanism of natural resource assets, which is a key part of China's ecological civilization construction. On July 6, 2018, an evaluation and acceptance meeting for the pilot programs concerning the unified confirmation and registration of natural resources was held in Beijing by seven ministries and commissions, including the Ministry of Natural Resources. At the meeting, pilot programs of several provinces, municipalities and autonomous regions passed the acceptance, indicating that much progress has been made in the field of confirmation and registration of natural resources after over one year's hard work. The state also focused on exploring the confirmation and registration of national parks, wetlands, water flows, proven reserves of mineral resources. On the basis of real estate registration, with the core mission being making a clear distinction between national ownership and collective ownership, between national ownership and governments at different levels assuming ownership, between different collective owners, and between different types of natural resources, and bearing in mind the goal of adopting a holistic approach to conserving our mountains, rivers, forests, farmlands, lakes and grasslands, local governments completed the work on investigating resource ownership, establishing registration units, confirming and registering, constructing databases, etc., resulting an effective set of workflow, technical methods and specifications.

Annex 4: Evidence from Chapter 4

4.1 China's Policy Implementation for Biodiversity Conservation

In regard of biodiversity conservation policies, we've already had relevant legislation, technological innovation and international mechanisms in China. A preliminary legal framework for biodiversity conservation has been established and technological innovation and international collaboration are making continuous progress.

In China, we have the *Constitution*, the fundamental law of the state, the *Environmental Protection Law of the People's Republic of China*, the basis of the environmental law system and a set of separate laws and administrative regulations on biodiversity conservation issued on the spirit of the above laws, such as the *Marine Environment Protection Law*, *Water Law*, *Water Pollution Prevention and Control Law*, *Water and Soil Conservation Law*, *Fishery Law*, *Forest Law*, *Grassland Law*, *Wild Animal Conservation Law*, *Regulations on Wild Plants Protection*, *Regulations on the Protection of Terrestrial Wild Animals*, *Regulations on the Protection of New Varieties of Plants*, *Regulations on Nature Reserves* and *Regulations on the Administration of Scenic and Historic Areas*. Besides, we also have local regulations on biodiversity conservation, for instance, *the Regulations on the Protection of Wild*

Aquatic Animals, *the Aquatic Resources Breeding Protection Regulations*, *Law on the Exclusive Economic Zone and the Continental Shelf* and *the Regulations on the Protection of Fishery Resources Breeding of Bohai*. Administrative regulations on biodiversity conservation in wetlands include the *Ramsar Convention*, *Convention on Biological Diversity*, etc.

In terms of local legislation, 9 provinces have established relevant regulations, including the *Regulations on the Protection of Wetlands in Heilongjiang Province*, *Regulations on the Protection of Wetlands in Gansu Province* and *Regulations on the Protection of Wetlands in Poyang Lake in Jiangxi Province*. In addition, a series of administrative laws and regulations have been issued, including regulations on nature reserves, regulations on the protection of wild plants, regulations on the safety management of agricultural GMOs, regulations on the administration of the import and export of endangered wild flora and fauna, regulations on the protection of wild medicine resources, etc. Some provincial governments and relevant authorities in charge have also formulated corresponding rules and regulations.

China has joined several international conventions related to biodiversity conservation, including the *Convention on Biological Diversity*, *Convention on Wetlands of International Importance Especially as Waterfowl Habitat*, *Convention on International Trade in Endangered Species of Wild Fauna and Flora*, *Convention Concerning the Protection of the World Cultural and Natural Heritage*, *Declaration of the United Nations Conference on the Human Environment* and *Rio Declaration on Environment and Development*. Laws concerning the management of introduced species include the *Law on the Entry and Exit Animal and Plant Quarantine*, *Animal Epidemic Prevention Law*, *Marine Environment Protection Law*, *Regulations on the Prevention of Livestock Epidemics*, etc. As for the emerging safety issues concerning GMOs, the State Council has issued the *Regulations on Administration of Agricultural Genetically Modified Organisms Safety* in 2001. The promulgation of those laws and regulations has efficiently supervised and promoted the conservation of biodiversity in China.

The Supreme People's Court of China has set up a division for environmental resources and issued guidelines on conducting specialized investigations and trials of biodiversity conservation-related cases, so as to guide courts at all levels to classify cases based on different basins or eco-function areas, unify judicial criteria, and improve the multiple-channel dispute settlement mechanism, thus laying a solid foundation for enhancing the juridical protection of environmental resources including biodiversity. Chinese courts give full play to the role of environmental public litigation, trying public interest litigation cases concerning wetlands, forestry, endangered plants, migratory birds in accordance with relevant laws. In the ancient Wucheng Town of Yongxiu County near Poyang Lake, the first biodiversity judicial protection base has been established. Adhering to modern judicial concepts such as strict law enforcement, safeguarding rights and interests, focusing on prevention and restoration and encouraging public participation, the base aims to make the best of judicial services in the process of advancing ecological civilization construction through circuit courts and legal publicity.

Basic surveys, scientific researches and monitoring of biodiversity have been conducted and technological innovation has been applied to promote the sustainable development of biodiversity. Relevant departments have organized a series of national and regional surveys, researches and monitoring on species and established corresponding databases and they have published several species catalogues such as the *Flora of China*, *Fauna of China*, *Cryptogamia of China*, *China Red Data Book of* Endangered Animals, etc. China has also drawn on international advanced experience and carried out demonstration projects, strengthened researches on the evaluation and management system of biological genetic resources and tried to build a mechanism to communicate relevant traditional knowledge and share benefits, thus coordinating the relationship among knowledge protection, expansion and utilization.

China has raised public awareness to participate and strengthened international cooperation and exchanges. Publicity campaigns on biodiversity conservation in various forms have been launched and education in this regard has also been enhanced in the campus. Public monitoring and reporting systems for biodiversity conservation have been established and improved. Partnerships on biodiversity conservation have been built in order to give full play to the role of non-governmental non-profit organizations and philanthropic organizations and mobilize stakeholders both in and out of China to promote the sustainable use of biodiversity resources. Moreover, China always sticks to its commitment to those conventions, introduces advanced experience from abroad and actively participates in formulating relevant international rules.

4.2 Additional Major Institutions with Conservation Management Responsibilities in China

In 2010, the General Assembly of the United Nations declared 2011-2020 the United Nations Decade on Biodiversity. The State Council established the *National Committee for 2010 International Year of Biodiversity* and held a meeting on which they passed *the China Action Plan for 2010 International Year of Biodiversity and China National Biodiversity Conservation Strategy and Action Plan* (2011-2030). In the June of 2011, the State Council decided to change the name of the Committee to "China National Committee for Biodiversity Conservation", it will continue to coordinate the efforts to protect biodiversity and guide China's action plan for the UN Decade on Biodiversity.

In 1992, the Biodiversity Committee of the Chinese Academy of Sciences (BC-CAS) was established to coordinate researches on biodiversity. Its responsibilities are as follows: to make biodiversity research policies of CAS; to make a long-term guideline and work plan for CAS's biodiversity researches; to review the rules and regulations on observation and experiments, organizational management mechanisms and fund allocation plans; to inspect the utilization of funds and the performance of work; to review academic exchanges and training programs; to make plans for domestic and international collaborative researches. BC-CAS will strive to implement the sub-project of "*Biodiversity Research and Information Management*", an environmental technical assistance project with loans from the World Bank. So far, 30+ databases have been established, 25 of which contain over 140,000 records that can be accessed via Internet.

4.3 DFIs Governance Structures for Conservation

● European Bank for Reconstruction and Development (EBRD): Borrowers are tasked with overseeing all management, monitoring and reporting.

● International Finance Corporation (IFC): IFC works with private-sector borrowers, creating a triangular oversight relationship: IFC, client and client's national government. Clients are generally tasked with monitoring and reporting, except for situations where national governments have domain over a natural resource or oversight responsibilities. In complex or high-risk scenarios, clients will be required to use the services of outside experts.

● Asian Infrastructure Investment Bank (AIIB) and Development Bank of Latin America (CAF): Borrowers are tasked with monitoring and reporting. The DFI may also carry out periodic site visits and works with implementers to mitigate any harm that has been caused.

● KFW agrees to a monitoring and reporting plan with the borrower or client, who is then empowered to manage that plan.

● Asian Development Bank (ADB): Borrowers compile regular reports, while the ADB maintains responsibility for due diligence in reviewing these reports. The ADB also carries out periodic site visits and works with implementers to mitigate any harm that have been caused.

● African Development Bank (AFDB): The AFDB will occasionally carry out independent audits of projects with substantial risks to biodiversity, including the use of third-party auditors. In cases where problems come to light, it designs action plans with measurable outcomes in conjunction with the borrower, with the aim of strengthening local capacity to monitor and manage projects and mitigate harm.

● Inter-American Development Bank (IDB) and World Bank (WB): These DFIs monitor compliance and oversee reporting.

Table A4-1 Commonalities Among DFI Guidelines for Project-Level Grievance Mechanisms

	AFDB	ADB	AIIB	EBRD	EIB	IFC	KFW	WB
Institutional Location								
It should be independent and monitored by a 3rd party	X							
It may be internal or external, as the DFI deems suitable			X					
Resources								
It should be scaled to the risks and impacts of the project		X	X	X	X	X	X	X
It should be adequately budgeted and staffed					X			
Design and establishment								
It should be designed in cooperation with the borrower/client to ensure legitimacy, accessibility, predictability and equitability	X							
It should be established as early as possible in the project development process				X				
Process								
It should address affected people's concerns promptly	X	X		X	X	X		
It should use a clear and transparent process		X	X	X	X	X		
It should have a predictable process	X				X			
It should be gender responsive or sensitive		X	X					
It should be culturally appropriate		X	X	X		X	X	
It should be free from manipulation, coercion or interference				X				
It should have a publicly accessible register of cases and outcomes	X		X					
It should report regularly to the public on its implementation				X	X			

ANNEXS

	AFDB	ADB	AIIB	EBRD	EIB	IFC	KFW	WB
Treatment of complainants								
It should protect complainants from intimidation/retaliation			X	X		X		
It should allow complainants to be remain anonymous if requested			X		X			
It should be free of cost to stakeholders	X				X	X		
It should be readily accessible to all segments of affected people			X	X				
The client should inform stakeholders of its availability			X	X		X		

Note: AFDB: African Development Bank; ADB: Asian Development Bank; AIIB: Asian Infrastructure Investment Bank; EBRD: European Bank for Reconstruction and Development; EIB: European Investment Bank; IFC: International Finance Corporation; KFW: German development bank, originally Kreditanstalt für Wiederaufbau; WB: World Bank.

参考文献

[1] PETRI PETER, MICHAEL PLUMMER. The Economic Effects of the Trans-Pacific Partnership：New Estimates[R]. Washington，DC：Peterson Institute for International Economics，2016.

[2] 中国一带一路网. 图解："一带一路"倡议六年成绩单[EB/OL]. [2019-09-09]. https：//www.yidaiyilu.gov.cn/xwzx/gnxw/102792.htm.

[3] World Bank. Belt and Road Economics：Opportunities and Risks of Transport Corridors[R]. Washington，D.C：World Bank，2019.

[4] 中国商务部. "一带一路"经贸合作取得新发展新提高新突破[EB/OL]. [2020-01-09]. http：//www.mofcom.gov. cn/article/ae/ai/202001/20200102928961.shtml.

[5] BHATTACHARYA A，GALLAGHER K P，MUÑOZ CABRÉ M，et al. Aligning G20 Infrastructure Investment with Climate Goals and the 2030 Agenda[R]. Foundations 20 Platform，a report to the G20，2019.

[6] YOSHINO NAOYUKI, ABIDHADJAEV UMID. Impact of Infrastructure Investment on Tax：Estimating Spillover Effects of the Kyushu High-Speed Rail Line in Japan on Regional Tax Revenue[R]. ADBI Working Papers 574，Asian Development Bank Institute，2016.

[7] DREHER，AXEL，FUCHS A，et al. Aid，China，and Growth：Evidence from a New Global Development Finance Dataset[R]. AidData Working Paper #46. Williamsburg，VA：AidData at William & Mary，2017.

[8] ASCENSÃO F，FAHRIG L，CLEVENGER A P，et al. Environmental challenges for the Belt and Road Initiative[J]. Nature Sustainability，2018，1：206-209.

[9] HUGHES，ALICE. Understanding and minimizing environmental impacts of the Belt and Road Initiative[J]. Conservation Biology，2019，33（4）：883-894.

[10] LOSOS, CLAIRE E，PFAFF，et al. Reducing Environmental Risks from Belt and Road Initiative Investments in Transportation Infrastructure（English）[R]. Policy Research working

paper; no. WPS 8718. Washington, D.C.: World Bank Group, 2019.

[11] NARAIN, DIVYA, MARON M, et al. Best-Practice Biodiversity Safeguards for Belt and Road Initiative's Financiers[J]. Nature Sustainability, 2020, 3 (8): 1-8.

[12] COSTANZA R, et al. Changes in the global value of ecosystem services[J]. Global Environmental Change, 2014, 26: 152-158.

[13] DAMANIA, RICHARD, DESBUREAUX S, et al. When Good Conservation Becomes Good Economics[R]. Washington, DC: World Bank, 2019.

[14] Global Environment Facility. Mainstreaming Gender at the GEF[R]. Washington, D.C.: GEF, 2013.

[15] ROCHELEAU, DIANNE E. Gender and Biodiversity: A Feminist Political Ecology Perspective[R]. Institute of Development Studies Bulletin, 1995, 26 (1): 9-16.

[16] CABRERA R, ISIDRO, MARTELO E Z, et al. Gender, Rural Households, and Biodiversity in Native Mexico[J]. Agriculture and Human Values, 2001, 18: 85-93.

[17] World Bank. Gender in Agriculture Sourcebook[R]. Washington, D.C.: World Bank, 2009.

[18] LU H N, LIANG X H, WANG C P. Improving Gender Equality Through China's Belt and Road Initiative[R]. Beijing and London: British Council, 2018.

[19] AGARWAL, BINA. Participatory Exclusions, Community Forestry, and Gender: An Analysis for South Asia and a Conceptual Framework[J]. World Development, 2001, 29 (10): 1623-1648.

[20] CORNWALL, ANDREA. Whose Voices? Whose Choices? Reflections on Gender and Participatory Development[J]. World Development, 2003, 31 (8): 1325-1342.

[21] MOSER, CAROLINE. Gender Planning and Development: Theory, Practice and Training[M]. London: Routledge. Biodiversity Finance Initiative (BIOFIN). United Nations Development Programme, 1993.

[22] XI JINPING. "Work Together to Build the Silk Road Economic Belt and The 21st Century Maritime Silk Road." Opening Ceremony Speech of the Belt and Road Forum for International Cooperation[J]. Quishi Journal, 14 May, 2017, 9 (3): 32.

[23] World Bank. Safeguards and Sustainability Policies in a Changing World[R]. Independent Evaluation Group. Washington, D.C.: World Bank, 2010.

[24] IRWIN, AMOS, KEVIN P. GALLAGHER Chinese Mining in Latin America: A Comparative Perspective[J]. Journal of Environment and Development, 2013, 22 (2): 207-234.

[25] RAY, REBECCA, GALLAGHER K P, et al. China in Latin America: Lessons for South South Cooperation for Sustainable Development[R]. Global Development Policy Center, Boston University, 2015.

[26] World Bank. Annual Report 2020: Ending Poverty, Investing in Opportunity[R]. Washington, DC: World Bank, 2020.

[27] RAY, REBECCA, GALLAGHER K P, et al. Development Banks and Sustainability in the Andean Amazon[M]. London: Routledge, 2019.

[28] Asian Infrastructure Investment Bank (AIIB). Environmental and Social Framework[EB/OL]. Beijing, AIIB, 2019. https://www.aiib.org/en/policies-strategies/framework-agreements/environmental-social-framework.html.

[29] African Development Bank (AfDB). Integrated Safeguard System: Policy Statement and Operational Safeguards[R]. Tunis: African Development Bank Group, 2013.

[30] World Bank. ESS6: Biodiversity Conservation and Sustainable Management of Living Natural Resources[R]. Washington DC.: World Bank, 2018.

[31] Development Bank of Latin America (CAF). Environmental and Social Safeguards for CAF/GEF Projects Manual[EB/OL]. Bogata, CAF, 2015.

[32] Convention on Biological Diversity. 2015-2020 Gender Action Plan[EB/OL]. Montreal: CBD. [2017-10-02]. https://www.cbd.int/gender/action-plan/.

[33] Climate Investment Funds. CIF Gender Action Plan - Phase 2[EB/OL]. Washington, D.C.: CIF. [2016-11-22]. https://www.climateinvestmentfunds.org/sites/default/files/ctf_scf_decision_by_mail_cif_gender_action_plan_phase_2_final_revised.pdf.

[34] Green Climate Fund. Mainstreaming Gender in Green Climate Fund Projects[R]. Yeonsu-gu, South Korea: GCF, 2017.

[35] International Finance Corporation. IFC Performance Standards[R]. Washington, D.C.: IFC, 2012.

[36] KVAM, REIDAR. Meaningful Stakeholder Engagement: A Joint Publication of the MFI Working Group on Environmental and Social Standards[R]. Washington, DC: Inter-American

Development Bank, 2019.

[37] International Union for the Conservation of Nature(IUCN), World Resources Institute(WRI). A Guide to the Restoration Opportunities Assessment Methodology[R]. Gland, Switzerland: IUCN, 2014.

[38] GUIDO SCHMIDT-TRAUB. Learning from China to protect nature[N]. China Dialogue, 2020-03-24.

[39] MONTANARELLA, LUCA, SCHOLES R, et al. The IPBES assessment report on land degradation and restoration[R]. Bonn, Germany: Intergovernmental Science-Policy Platform on Biodiversity and Ecosystem Services, 2018.

[40] 刘桂环. 探索中国特色生态补偿制度体系[N]. 中国环境报, 2019-12-18.

[41] 董战峰, 李红祥, 葛察忠, 等. 环境经济政策年度报告2017[J]. 环境经济, 2018, 4: 12-35.

[42] 大自然保护协会. 中国TNC与浙江龙坞合作开展水源地保护项目[EB/OL]. http://www.tnc.org.cn/#News#schedule#iframe99dc279553caa331d70c9f0840779587b1f0c4fddb7a32175cd9319c7a817b5db938ef981a6ed605397fb1.

[43] 万向信托. 创新业务新模式万向信托推出全国首个水基金信托[EB/OL]. http://biz.zjol.com.cn/system/2015/11/18/020917870.shtml.

[44] SCHOMERS, SARAH, MATZDORF B. Payments for Ecosystem Services: A Review and Comparison of Developing and Industrialized Countries[J]. Ecosystem Services 6 December, 2013, 16-30.

[45] SALZMAN, JAMES, BENNETT G, et al. The Global Status and Trends of Payments for Ecosystem Services[J]. Nature Sustainability, 2018, 1: 136-144.

[46] BULL, JOSEPH W, SUTTLE K B, et al. Biodiversity Offsets in Theory and Practice[J]. Oryx 47: 3 July, 2013: 369-380.

[47] GELCICH, STEFAN, VARGAS C, et al. Achieving Biodiversity Benefits with Offsets: Research Gaps, Challenges, and Needs[J]. Ambio, 2017, 46: 184-189.

[48] MCKENNEY, BRUCE A, KIESECKER J M. Policy Development for Biodiversity Offsets: A Review of Offset Frameworks[J]. Environmental Management, 2010, 45: 165-176.

[49] BULL, JOSEPH W, STRANGE N. The Global Extent of Biodiversity Offset Implementation under no Net Loss Policies[J]. Nature Sustainability, 2018, 1: 790-798.

[50] GARDNER, TOBY A, HASE A, et al. Biodiversity Offsets and the Challenge of Achieving No Net Loss[J]. Conservation Biology, 2013, 27（6）: 1254-1264.

[51] BEZOMBES, LUCIE, GAUCHERAND S, et al. Ecological Equivalence Assessment Methods: What Trade-Offs between Operationality, Scientific Basis and Comprehensiveness? [J]. Environmental Management, 2017, 60: 216-230.

[52] QUÉTIER, FABIEN, LAVOREL S. Assessing Ecological Equivalence in Biodiversity Offset Schemes: Key Issues and Solutions[J]. Biological Conservation, 2011, 144（12）: 2991-2999.

[53] LUCK G W, CHAN K M A, FAY J P. Protecting ecosystem services and biodiversity in the world's watersheds[J]. Conservation Letters, 2009, 2: 178-188.

[54] BURIAN, GABRIELA, SEALE J, et al. Business Case for Investing in Soil Health[R]. Geneva: World Business Council for Sustainable Development, 2018.

[55] CLARK, ROBYN, JAMES REED, et al. Bridging funding gaps for climate and sustainable development: Pitfalls, progress and potential of private finance. Land Use Policy, 2017, 71: 335-346. https: //doi.org/10.1016/j.landusepol. 2017.12.013.

[56] BASS, MARGOT, FINER M, et al. Global Conservation Significance of Ecuador's Yasuní National Park[J]. PLOS One, 2010, 5（1）: e8767.

[57] KLINGER, MICHELLE J. In Their Own Time, on Their Own Terms: Improving development bank project outcomes through community-centered sustainable development partnerships in the Brazilian Amazon[R]. Boston University Global Development Policy Center, 2019.

[58] DILLENBECK, MARK. National Environmental Funds: A New Mechanism for Conservation Finance[J]. Parks, 1994, 4（2）: 39-46.

[59] KNOX, JOHN H. The Neglected Lessons of the NAFTA Environmental Regime[J]. Wake Forest Law Review, 2010, 45: 391-424.

[60] North American Development Bank. North American Development Bank Annual Report[R]. San Antonio, TX: NAD Bank, 2019.

[61] EHLER, CHARLES, DOUVERE F. Marine spatial planning: A step-by-step approach toward ecosystem-based management[R]. Paris: UNESCO, 2009.

[62] Asian Development Bank（ADB）. Building Gender into Climate Finance: ADB Experience with Climate Investment Funds[R]. Manila: ADB, 2016.